ADMIRAL DAVID G. FARRAGUT

MAKERS OF AMERICAN HISTORY

ADMIRAL FARRAGUT

BY

CAPTAIN A. T. MAHAN, U. S. N.

President of the United States Naval War College, Author of "The Gulf and Inland Waters," and "The Influence of Sea Power Upon History, 1660-1783"

J. A. HILL & COMPANY

NEW YORK 1904

PREFACE.

In preparing this brief sketch of the most celebrated of our naval heroes, the author has been aided by the very full and valuable biography published in 1878 by his son, Mr. Loyall Farragut, who has also kindly supplied for this work many additional details of interest from the Admiral's journals and correspondence, and from other memoranda. For the public events connected with Farragut's career, either directly or indirectly, recourse has been had to the official papers, as well as to the general biographical and historical literature bearing upon the war, which each succeeding year brings forth in books or magazines. The author has also to express his thanks to Rear-Admiral Thornton A. Jenkins, formerly chief-of-staff to Admiral Farragut; to Captain John Crittenden Watson, formerly his flag-lieutenant; and to his friend General James Grant Wilson, for interesting anecdotes and reminiscences.

A. T. M.

CONTENTS.

LIST OF ILLUSTRATIONS.

ADMIRAL FARRAGUT.

CHAPTER I.

FAMILY AND EARLY LIFE.

1801–1811.

THE father of Admiral Farragut, George Farragut, was of unmixed Spanish descent, having been born on the 29th of September, 1755, in the island of Minorca, one of the Balearic group, where the family had been prominent for centuries. One of his ancestors, Don Pedro Ferragut, served with great distinction under James I, King of Aragon, in the wars against the Moors, which resulted in their expulsion from Majorca in 1229, and from the kingdom of Valencia, in the Spanish Peninsula, in 1238. As Minorca in 1755 was a possession of the British Crown, to which it had been ceded in 1713 by the Treaty of Utrecht, George Farragut was born under the British flag; but in the following year a French expedition, fitted out in Toulon, succeeding in wresting from the hands of Great Britain both the island and its excellent fortified harbor, Port Mahon, one of the most advantageous naval stations in the Mediterranean. It was in the course of the operations which resulted in this conquest of Minorca by the French that the British fleet, under the command of Admiral Byng, met with the check for which the admiral paid the penalty of his life a few months later. At the close of the Seven Years' War, in 1763, the island was restored to Great Britain, in whose hands it remained until 1782, when it was again retaken by the French and Spaniards.

George Farragut, however, had long before severed his connection with his native country. In March, 1776, he emigrated to North America, which was then in the early throes of the Revolutionary struggle. Having grown to manhood a subject to Great Britain, but alien in race and feeling, he naturally espoused the cause of the colonists, and served gallantly in the war. At its end he found himself, like the greater part of his adopted countrymen, called to the task of building up his own fortunes, neglected during its continuance; and, by so doing, to help in restoring prosperity to the new nation. A temper naturally adventurous led him to the border lines of civilization; and it was there, in the region where North Carolina and eastern Tennessee meet, that the years succeeding the Revolution appear mainly to have been passed. It was there also that he met and married his wife, Elizabeth Shine, a native of Dobbs County, North Carolina, where she was born on the 7th of June, 1765. At the time of their marriage the country where they lived was little more than a wilderness, still infested by Indians; and one of the earliest recollections of the future admiral was being sent into the loft, on the approach of a party of these, while his mother with an axe guarded the door, which she had barricaded. This unsettled and dangerous condition necessitated a constant state of preparedness, with some organization of the local militia, among whom George Faragut held the rank of a major of cavalry, in which capacity he served actively for some time.

While resident in Tennessee, George Farragut became known to Mr. W. C. C. Claiborne, at that time the member for Tennessee in the National House of Representatives. Mr. Claiborne in 1801 became governor of Mississippi Territory; and in 1803, when the United States purchased from France the great region west of the Mississippi River, to which the name Louisiana was then applied, he received the cession of the newly acquired possession. This was soon after divided into two parts by a line following the thirty-third parallel of north latitude, and Claiborne became governor of the southern division, which was called the Territory of Orleans. To this may probably be attributed the removal of the Farraguts to Louisiana from eastern Tennessee. The region in which the latter is situated, remote both from tide-water and from the great river by which the Western States found their way to the Gulf of Mexico, was singularly unfitted to progress under the conditions of communication in that day; and it long remained among the most backward and primitive portions of the United States. The admiral's father, after his long experience there, must have seen that there was little hope of bettering his fortunes. Whatever the cause, he moved to Louisiana in the early years of the century, and settled his family in New Orleans. He himself received the appointment of sailing-master in the navy, and was ordered to command a gun-boat employed in the river and on the adjacent sounds. A dispute had arisen between the United States and the Spanish Government, to whom the Floridas then belonged, as to the line of demarcation between the two territories; and George Farragut was at times employed with his vessel in composing disturbances and forwarding the views of his own government.

David Glasgow, the second son of George Farragut, and the future Admiral of the United States Navy, was born before the removal to Louisiana, on the 5th of July, 1801, at Campbell's Station, near Knoxville, in eastern Tennessee. In 1808, while living in his father's house on the banks of Lake Pontchartrain, an incident occurred which led directly to his entrance into the navy, and at the same time brought into curious coincidence two families, not before closely associated, whose names are now among the most conspicuous of those in the annals of the navy. While George Farragut was fishing one day on Lake Pontchartrain he fell in with a boat, also engaged in fishing, in which was an old gentleman prostrated by the heat of the sun. He took him to his own house, where he was cared for and nursed until he died, never having recovered strength sufficient to be removed. The sufferer was David Porter, the father of the Captain David Porter who afterward commanded the frigate Essex in her adventurous and celebrated cruise in the Pacific during the years 1813 and 1814, and grandfather of the still more distinguished Admiral David D. Porter, who, over half a century later, served with David Farragut on the Mississippi in the civil war, and in the end succeeded him as second admiral of the navy. Captain, or rather, as he then was, Commander Porter being in charge of the naval station at New Orleans, his father, who had served actively afloat during the Revolution and had afterward been appointed by Washington a sailing-master in the navy, had obtained orders to the same station, in order to be with, though nominally under, his son. The latter deeply felt the kindness shown to his father by the Farraguts. Mrs. Farragut herself died of yellow fever, toward the end of Mr. Porter's illness, the funeral of the two taking place on the same day; and Commander Porter soon after visited the family at their home and offered to adopt one of the children. Young David Farragut then knew little of the element upon which his future life was to be passed; but, dazzled by the commander's uniform and by that of his own elder brother William, who had received a midshipman's warrant a short time before, he promptly decided to accept an offer which held forth to him the same brilliant prospects. The arrangement was soon concluded. Porter promised to be to him always a friend and guardian; and the admiral wrote in after life, "I am happy to have it in my power to say, with feelings of the warmest gratitude, that he ever was to me all that he promised." The boy returned to New Orleans with his new protector, in whose house he thenceforth resided, making occasional trips across Lake Pontchartrain to a plantation which his father had purchased on the Pascagoula River. A few months later Commander Porter appears to have made a visit to Washington on business connected with the New Orleans station, and to have taken Farragut with him to be placed at school, for which there were few advantages at that time in Louisiana. The boy then took what proved to be a last farewell of his father. George Farragut continued to live in Pascagoula, and there he died on the 4th of June, 1817, in his sixty-second year.

The trip north was made by Porter and his ward in the bomb-ketch Vesuvius, a stop being made at Havana; where the commander had business growing out of the seizure by him in the Mississippi River of some French privateers, for which both Spain and the United States had offered a reward. At Havana the lad heard of an incident, only too common in those days, which set his heart, as those of his countrymen were fast being set, against Great Britain. Presuming confidently upon the naval weakness of the United States, and arguing from their long forbearance that insults to the flag would be indefinitely borne for the sake of the profitable commerce which neutrality insured, Great Britain, in order to support the deadly struggle in which she was engaged with France, had endeavored to shut off the intercourse of her enemy with the rest of the world, by imposing upon neutral trade restrictions before unheard of and without justification in accepted international law. Both the justice and policy of these restrictions were contested by a large party of distinguished Englishmen; but upon another principle men of all parties in the old country were practically agreed, and that was the right of the British Government to compel the services of British seamen wherever found. From this grew the claim, which few Englishmen then dared to disavow, that their ships of war could rightfully take from any neutral merchant ship any seaman of British birth who was found on board. In estimating this monstrous pretention, Americans have shown little willingness to allow for the desperate struggle in which Great Britain was involved, and the injury which she suffered from the number of her seamen who, to escape impressment in their home ports and the confinement of ships of war, sought service in neutral merchant ships. Her salvation depended upon her navy; and seamen were so scarce as seriously to injure its efficiency and threaten paralysis. This was naturally no concern of the United States, which set up its simple, undeniable right to the protection the neutral flag should give to all persons and goods under it, which were not involved in any infraction of belligerent rights. The straits of Great Britain, however, were too dire to allow the voice of justice to override that of expediency. Had the United States Navy been a force as respectable in numbers as it was in efficiency, the same dictates of expediency might have materially controlled the action of her opponent; might have prevented outrage and averted war. As it was, right was set up against right—the right of the neutral flag on the one hand against the right of a country to the service of all her citizens on the other. The United States protested and wrote with all the conviction of a state upon whose side justice was. She resorted to measure after measure of peaceable coercion; but she had no military force to show upon the sea, and her utterances were consequently too uncertain to command respect. Great Britain continued to take seamen from American merchant ships upon the plea of her right to impress British seamen in any place; and, though the claim to detain or search ships of war had been explicitly disavowed after the Chesapeake affair of 1807, scant deference was shown to the vessels of a power so little able to stand up for itself. In a day when most vessels carried some guns for self-defense, it was a simple matter to ignore the national character of an armed ship and to stop it unceremoniously. Of such an insult Farragut heard during this stay in Havana. The brig Vixen, of the United States Navy, had been fired into by a British ship of war. "This," wrote Farragut in his journal, "was the first thing that caused in me bad feeling toward the English nation. I was too young to know anything about the Revolution; but I looked upon this as an insult to be paid in kind, and was anxious to discharge the debt with interest." It is scarcely necessary to say how keenly this feeling was shared by his seniors in the service, to whom the Vixen incident was but one among many bitter wrongs which the policy of their Government had forced them humbly to swallow.

After their arrival in Washington Farragut was put to school, where he remained until Porter was relieved from the New Orleans station. During his stay at the capital he was presented by his guardian to the Secretary of the Navy, Paul Hamilton, of South Carolina; who, after ascertaining his wish to enter the service, promised him a midshipman's warrant when he should be ten years old. The promise was more than kept, for the warrant, when issued, was dated December 17, 1810; the future admiral thus finding himself at least a titular officer, in the service which he was afterward to adorn, when not quite nine and a half years of age. Although at that time, and in earlier generations, boys, no older than Farragut then was, were not infrequently turned aboard ship to fight their own way in life, Porter did not so construe his duties to his charge. In the latter part of 1810 he finally left New Orleans and went

North again, this time by the Mississippi River and in a gun-boat. The voyage to Pittsburg against the swift current took three months; and it was not till toward the close of the year that he and his family were again settled in their home at Chester, in Pennsylvania, the birthplace of Mrs. Porter. Farragut was then removed from Washington and put to school in Chester, there to remain until his guardian should be able to take him to sea under his own eyes, in a vessel commanded by himself. This opportunity was not long in arriving.

CHAPTER II.

CRUISE OF THE ESSEX.

1811–1814.

CHILD though Farragut was when he obtained his nominal admission to the navy, he had but a short time to wait before entering upon its stern realities —realities far harsher in that day than now. The difficulties that had existed between the United States and Great Britain, ever since the outbreak of war between the latter and France in 1793, were now fast drifting both nations to the collision of 1812. The Non-intercourse Act of March, 1809, forbidding American merchant ships to enter any port of France or Great Britain, as a retaliation for the outrages inflicted by both upon American commerce, had expired by its own limitations in May, 1810, when commerce with the two countries resumed its natural course; but Congress had then passed a proviso to the effect that if either power should, before March 3, 1811, recall its offensive measures, the former act should, within three months of such revocation, revive against the one that maintained its edicts. Napoleon had contrived to satisfy the United States Government that his celebrated Berlin and Milan decrees had been recalled on the 1st of November; and, consequently, non-intercourse with Great Britain was again proclaimed in February, 1811. The immediate result was that two British frigates took their station off New York, where they overhauled all merchant ships, capturing those bound to ports of the French Empire, and impressing any members of the crews considered to be British subjects. The United States then fitted out a squadron, to be commanded by Commodore John Rodgers; whose orders, dated May 6, 1811, were to cruise off the coast and to protect American commerce from unlawful interference by British and French cruisers. Ten days later occurred the collision between the commodore's ship, the President, and the British corvette Little Belt. Of Rodgers's squadron the frigate Essex, expected shortly to arrive from Europe, was to be one; and Commander Porter, who did not obtain his promotion to the grade of captain until the following year, was ordered to commission her. He took his ward with him, and the two joined the ship at Norfolk, Virginia, in August, 1811, when the young midshipman had just passed his tenth birthday. Long years afterward Mrs. Farragut was told by Commodore Bolton, one of the lieutenants of the Essex, that he remembered to have found the little boy overcome with sleep upon his watch, leaning against a gun-carriage, and had covered him with his pea-jacket to protect him from the night air. An amusing incident, however, which occurred during these first months of his naval career showed that the spirit of battle was already stirring. Porter, probably with a view to keep the lad more immediately under his own eye, had made him midshipman of his gig, as the captain's special boat is called. On one occasion he was sent in to the wharf, to wait for the captain and bring him to the ship when he came. A crowd of dock-loungers gradually collected, and the youngster who stood erect in the boat, doubtless looking pleasedly conscious of his new uniform and importance, became the object of audible comment upon his personal appearance. The boat's crew sat silent but chafing, the bowman holding on with his boat-hook, until one loafer proceeded from witticism to practical joking by sprinkling the midshipman with an old water-pot. Quick as look the bowman caught his boot-hook in the culprit's pocket and dragged him into the boat, while the rest of the crew, by this time spoiling for a fight, seized their stretchers, jumped ashore, and began laying on right and left. Farragut, so far from restraining, went with them, waving his dirk and cheering them on. The victorious seamen fought their way up to Market Square, where the police interfered, arresting all parties, and the little officer was formally bound over to keep the peace.

The Hartford, upon which Farragut first hoisted his admiral's flag, has obtained a particular interest from its close association with the whole of his course of victory; and the Essex, a ship of very different type, would attract attention as the one that cradled his career, and witnessed the part of it which is only second in excitement to his exploits as a commander-in-chief, had she no special claims of her own to notice. But the Essex, both in her origin and through her subsequent history, especially when under Porter's command, was a marked ship. She was an offspring of the quarrel between the United States and the French Republic, which arose out of the extravagant demands made by the latter upon the compliance of her former ally, in consequence of the service which it was claimed had been rendered during the Revolutionary War. Ignoring the weakness of the American Republic, and the dependence of a large section of the country upon commerce, the French Government had expected that it should resist, even by force, the seizure by British cruisers of French property in American vessels, and thus bring on hostilities with Great Britain; and that, although the United States Government admitted the practice of capturing enemy's property in neutral ships, however objectionable in theory, to be part of the traditional and recognized law of nations. Going on from step to step, in the vain endeavor by some means to injure the maritime predominance of Great Britain, which defied the efforts both of their navy and of their privateers, the French Legislature in January, 1798, decreed that any neutral vessel which should be found to have on board, not merely British property, but property, to whomsoever belonging, which was grown or manufactured in England or her colonies, should be a lawful prize to French cruisers. This extravagant claim, which not only seized goods that had been heretofore and by all others accounted free, but also, contrary to precedent, confiscated the vessel as well as the cargo, broke down the patience of the United States, where the Government was then still in the hands of the Federalists, whose sympathies were rather British than French. Nearly a year before, President Adams had called a special meeting of Congress and recommended an increase of the navy, to the numerical weakness of which was due the recklessness with which both Great Britain and France inflicted insult and injury upon our seamen and upon our commerce. That the United States of that day, so inferior in wealth and numbers to both belligerents, should dream of entering the lists with either singly, was perhaps hopeless; but through the indifference of Congress the navy of a people, then second only to the English as maritime carriers, was left so utterly impotent that it counted for naught, even as an additional embarrassment to those with which the contending powers were already weighted. When, therefore, in retaliation for the seizures made under the French decree of January, 1798, Congress, without declaring war, directed the capture of French armed vessels, wherever found on the high seas, it became necessary to begin building a navy which to some slight degree might carry out the order. An act, intended to hasten the increase of the navy, was passed in June, 1798, authorizing the President to accept such vessels as might be built by the citizens for the national service, and to issue six-per-cent stock to indemnify the subscribers.

Under this law the Essex was built in Salem, Massachusetts, by a subscription raised among the citizens. As the project grew, and the amount likely to be obtained became manifest, the purpose to which it should be devoted was determined to be the building of a frigate of thirty-two guns; one of the well-recognized, but smaller, classes under which the vessels called frigates were subdivided. Except the work of the naval architect proper, the model and the superintendence, which were undertaken by a gentleman from Portsmouth, everything in the building and equipment was portioned out among Salem men, and was supplied from the resources of the town or of the surrounding country. During the winter of

1798 to 1799 the sleds of all the farmers in the neighborhood were employed bringing in the timber for the frames and planking of the new ship. The rigging was manufactured by the three rope-walks then in the place, each undertaking one mast; and the sails were of cloth so carefully selected and so admirably cut that it was noticed the frigate never again sailed so well as with this first suit. When the rope cables, which alone were then used by ships instead of the chains of the present day, were completed, the workmen took them upon their shoulders and marched with them in procession to the vessel, headed by a drum and fife. The building of the Essex was thus an effort of city pride and local patriotism; and the launch, which took place on the 30th of September, 1799, became an occasion of general rejoicing and holiday, witnessed by thousands of spectators and greeted by salutes from the battery and shipping. The new frigate measured 850 tons, and cost, independent of guns and stores, somewhat over $75,000. Her battery in her early history was composed of twenty-six long twelve-pounders on the main deck, with sixteen thirty-two-pound carronades and two chase guns on the deck above. At a later day, and during the cruise under Porter, this was changed to forty thirty-two-pound carronades and six long twelves. This battery, though throwing a heavier weight, was of shorter range than the former; and therefore, though advantageous to a ship able to choose her position, was a fatal source of weakness to a slow or crippled vessel, as was painfully apparent in the action where the Essex was lost.

Notwithstanding the zeal and emulation aroused by the appeal to Salem municipal pride, and notwithstanding the comparative rapidity with which ships could then be built, the Essex in her day illustrated the folly of deferring preparation until hostilities are at hand. The first French prize was taken in June, 1798, but it was not till December 22d of the following year that the Essex sailed out of Salem harbor, commanded then by Edward Preble, one of the most distinguished officers of the early American navy. Newport was her first port of arrival. From there she sailed again on the 6th of January, 1800, in company with the frigate Congress, both being bound for Batavia, whence they were to convoy home a fleet of merchant ships; for in the predatory warfare encouraged by the French Directory, the protection of our commerce from its cruisers was a duty even more important than the retaliatory action against the latter, to which the *quasi* war of 1798 was confined. When six days out, the Congress was dismasted. The Essex went on alone, and was thus the first ship-of-war to carry the flag of the United States around the Cape of Good Hope into the Indian Ocean. A dozen years later the bold resolution of Porter to take her alone and unsupported into the Pacific, during the cruise upon which young Farragut was now embarking, secured for this little frigate the singular distinction of being the first United States ship-of-war to double Cape Horn as well as that of Good Hope. In the intervening period the Essex had been usefully, but not conspicuously, employed in the Mediterranean in the operations against Tripoli and in protecting trade. In 1811, however, she was again an actor in an event of solemn significance. Upon her return to the United States, where Porter was waiting to take command, she bore as a passenger William Pinkney, the late Minister to Great Britain; who, after years of struggle, on his part both resolute and dignified, to obtain the just demands of the United States, had now formally broken off the diplomatic relations between the two powers and taken an unfriendly leave of the British Government

Being just returned from a foreign cruise, the Essex needed a certain amount of refitting before again going to sea under her new commander; but in October, 1811, she sailed for a short cruise on the coast, in furtherance of the Government's orders to Commodore Rodgers to protect American commerce from improper interference. Orders of such a character were likely at any moment to result in a collision, especially in the hands of a gallant, hasty officer scarcely out of his first youth; for Porter was at this time but thirty-one, and for years had felt, with the keen resentment of a military man, the passive submission to insult shown by Jefferson's government. No meeting, however, occurred; nor were the months that elapsed before the outbreak of war marked by any event of special interest except a narrow escape from shipwreck on Christmas eve, when the Essex nearly dragged on shore in a furious northeast gale under the cliffs at Newport. Farragut has left on record in his journal, with the proper pride of a midshipman in his ship, that the Essex was the smartest vessel in the squadron, and highly complimented as such by Commodore Rodgers. In acknowledgment of the skill and activity of his seamen, Porter divided the ship's company into three watches, instead of the usual two—an arrangement only possible when the smaller number in a watch is compensated by their greater individual efficiency. This arrangement continued throughout the cruise, until the ship was captured in 1814.

On the 18th of June, 1812, war was at last declared against Great Britain. The Essex had again been cruising during the spring months; but the serious character of the new duties before her made a thorough refit necessary, and she was not able to sail with the squadron under Commodore Rodgers, which put to sea from New York on the 21st of June. On the 3d of July, however, she got away, Porter having the day before received his promotion to post-captain, then the highest grade in the United States Navy. The ship cruised off the coast, making several prizes of vessels much inferior to herself in force, and on the 7th of September anchored within the capes of the Delaware. Much to Porter's surprise and annoyance, although ready to sail at once if furnished with provisions, none reached him. The ship was therefore taken up the Delaware and anchored off Chester, where she was prepared for a long and distant cruise directed against British commerce, the suggestion of which Porter believed came first from himself. By this a squadron consisting of the Constitution, Essex, and Hornet sloop-of-war, under the command of Commodore Bainbridge in the first-named frigate, were to proceed across the Atlantic to the Cape Verde Islands, thence to the South Atlantic in the neighborhood of Brazil, and finally to the Pacific, to destroy the British whale-fishery there. The plan was well conceived, and particularly was stamped with the essential mark of all successful commerce-destroying, the evasion of the enemy's cruisers; for, though the American cruisers were primed to fight, yet an action, even if successful, tended to cripple their powers of pursuit. A rapid transit through the Atlantic, with an ultimate destination to the then little-frequented Pacific, was admirably calculated to conceal for a long time the purposes of this commerce-destroying squadron. As it happened, both the Constitution and Hornet met and captured enemy's cruisers off the coast of Brazil, and then returned to the United States. Farragut thus lost the opportunity of sharing in any of the victories of 1812, to be a partaker in one of the most glorious of defeats.

The Constitution and Hornet being in Boston, and the Essex in the Delaware, it became necessary to appoint for the three a distant place of meeting, out of the usual cruising grounds of the enemy, in order that the ships, whose first object was to escape crippling, could pass rapidly through the belt of British cruisers then girding the coast of the United States. The brilliant record made by United States ships in their single combats with the enemy during this war should not be allowed to blind our people to the fact that, from their numerical inferiority, they were practically prisoners in their own ports; and, like other prisoners, had to break jail to gain freedom to act. The distant and little frequented Cape Verde group, off the African coast, was therefore designated as the first rendezvous for Bainbridge's squadron, and the lonely island of Fernando Noronha, off the coast of Brazil, close under the equator, as the second. Both of these places were then possessions of Portugal, the ally of Great Britain though neutral as to the United States. With these orders the Constitution and Hornet sailed from

Boston on the 26th of October, 1812, and the Essex two days later from the capes of the Delaware. Their course in the passage was to be so directed as to cross at the most favorable points the routes of British commerce.

On the 27th of November the Essex, after an uneventful voyage, anchored at Porto Praya, in the Cape Verdes, where she remained five days. Receiving no news of Bainbridge, Porter sailed again for Fernando Noronha. On the 11th of December a British packet, the Nocton, was captured, and from her was taken $55,000 in specie—an acquisition which contributed much to facilitate the distant cruise contemplated by Porter. Four days later the Essex was off Fernando Noronha, and sent a boat ashore, which returned with a letter addressed ostensibly to Sir James Yeo, of the British frigate Southampton; but between the lines, written in sympathetic ink, Porter found a message from Bainbridge, directing him to cruise off Rio and wait for the Constitution. On the 29th of December he was in the prescribed station, and cruised in the neighborhood for some days, although he knew a British ship-of-the-line, the Montagu, was lying in Rio; but only one British prize was taken, the merchant vessels of that nation usually waiting in port until they could sail under convoy of a ship-of-war. Attempting to get to windward in a heavy sea, the Essex was much racked and injured some of her spars, and her captain therefore decided to bear away for refit to St. Catherine's—a port five hundred miles south of Rio Janeiro, which had been named in his instructions as a contingent rendezvous. On the 20th of January, 1813, the Essex anchored there, and began the work of refitting and filling with water and fresh provisions. A few days after her arrival a small Portuguese vessel came in, bringing an account of the capture by the Montagu of an American corvette, which Porter supposed to be the Hornet, as well as a rumor of the action between the Constitution and the Java, and a report that re-enforcements were reaching the British naval force on the station. The history of past wars convinced Porter that the neutrality of the Portuguese port in which he was lying would not be respected by the enemy. In a very few days his presence there must become known; any junction with his consorts was rendered most unlikely by the news just received, and he determined at once to undertake alone the mission for which the three ships had been dispatched. With admirable promptitude, both of decision and action, the Essex sailed the same night for the Pacific.

From the time of leaving the United States the crew of the ship had been restricted to that close and economical allowance of provisions and water which was necessary to a vessel whose home ports were blocked by enemy's cruisers, and which in every quarter of the globe might expect to meet the fleets and influence of a powerful foe. The passage round Cape Horn, always stormy, was both a long and severe strain to a vessel bound from east to west, and dependent wholly upon sail; for the winds prevail from the westward. The utmost prudence was required in portioning out both food and water, and of bread there remained, on leaving St. Catherine's, only enough for three months at half allowance—that is, at half a pound per day. The boy Farragut thus found himself, at the outset of his career, exposed to one of the severest tests of his arduous calling—a long and stormy passage, made in the teeth of violent gales, and with a crew reduced to the scantiest possible allowance of food, under conditions when the system most demands support. In his journal he speaks, as Porter does in his, of the severe suffering and dreadful weather experienced. For twenty-one days the Essex struggled with the furious blasts, the heavy seas, and the bitter weather, which have made the passage round Cape Horn proverbial for hardship among seamen. On the 3d of March, he writes, a sea was shipped which burst in, on one side of the ship and from one end to the other, all the ports through which the guns are fired, and which, for such a passage, are closed and securely fastened. One boat on the weather side was driven in on deck, and that on the opposite carried overboard; but with great difficulty the latter was saved. Large quantities of water rushed below, leading those there to imagine that the ship was sinking. "This was the only instance in which I ever saw a real good seaman paralyzed by fear at the dangers of the sea. Several of the sailors were seen on their knees at prayer; but most were found ready to do their duty. They were called on deck, and came promptly, led by William Kingsbury, the boatswain's mate. Long shall I remember the cheering sound of his stentorian voice, which resembled the roaring of a lion rather than that of a human being, when he told them: 'D—n their eyes, to put their best foot forward, as there was one side of the ship left yet.'"

Cape Horn, however, was at last passed and enough ground gained to the westward to allow the Essex again to head north. On the 11th of March she was off the city of Valparaiso, in Chile. As far as Porter then knew, Chile was still a province of Spain, and Spain was the ally of Great Britain; whose armies for four years past had been engaged in war in the Peninsula, to shake from it the grip of Napoleon. There had been trouble also between Spain and the United States about the Floridas. The first lieutenant of the Essex was therefore first sent ashore to see what reception would be given, and returned with the satisfactory intelligence that Chile was in revolution against the mother country, and was ready heartily to welcome a ship-of-war belonging to the American Republic. He also brought the news that the Viceroy of Spain in Peru had fitted out privateers against Chilian commerce; and that these, on the plea of being allies of Great Britain, had begun to capture American whalers. It seemed, therefore, that the Essex had arrived as opportunely for the protection of United States interests as for the injury of British commerce.

Several days were lost in these preliminaries, so that it was not till the 15th that the anchor was dropped in Valparaiso. Despite the cordial reception given, Porter was in haste to reach his scene of action in the North and sailed again on the 22d. Four days later he met a Peruvian privateer, the Nereyda, the captain of which was deceived by the Essex hoisting British colors. Coming on board the frigate, he stated freely that the Spaniards considered themselves the allies of Great Britain, that he was himself cruising for American whalers, and had on board at the moment the crews of two of these which he had taken. Having extracted all the information he wanted, Porter undeceived the privateersman, took possession of the ship, threw overboard her guns and ammunition, and then released her, with a letter to the Viceroy; which, backed by the presence of the Essex, was calculated to insure peaceable treatment to American vessels.

There were at this time on the coast of Peru and in the neighboring waters twenty-three American whalers, worth, with their cargoes, two and a half million dollars, and mostly unarmed, having left home in a time of peace. Of English ships there were twenty; but, their country having been long at war, these were generally armed, and in many cases provided with letters of marque authorizing them to act as privateers and capture vessels hostile to their Crown. In this state of things, so unpromising for American interests, the arrival of the Essex entirely turned the scales, besides stopping the Spanish depredations which had but just begun.

On the 27th of March, off the harbor of Callao, the port of Lima, Porter recaptured the Barclay, one of the American ships seized by the Nereyda; but, although the frigate again disguised her nationality by hoisting British colors, there was among the several vessels in the harbor only one that showed the same flag. With the Barclay in company, the Essex now stood away for the Galapagos Islands. These are a group situated just south of the equator and some five hundred miles from the South American coast. Uninhabited then, as for the most part they still are, they were in 1813 a favorite rendezvous for British whalers, who had established upon one of the islands (Charles) a means of communication by a box nailed to a tree, which was called the post-office. They abound in turtle, some of which weigh several hundred pounds, and form a very valuable as well as acceptable change of diet to seamen long confined to salt food. On the 17th of April the Essex came in sight of Chatham Island, one of the largest, and remained cruising in the neighborhood of the group till the beginning of June, when want of water compelled her to go to Tumbez, a port on the continent just abreast of the Galapagos. In this period seven British whalers were taken; so that on the 24th of June there were anchored in Tumbez Bay, including the frigate and the Barclay, nine vessels under Porter's command. Of these, he commissioned one—the fastest and best, somewhat less than half the size of the Essex herself—as a United States cruiser, under his command. She was named the Essex Junior, carried twenty guns, of which half were long six-pounders and half eighteen-pounder carronades, and was manned by sixty of the Essex's crew under her first lieutenant.

The first service of the Essex Junior was to convoy to Valparaiso the Barclay and four of the British prizes. The occasion was one of great importance and interest to Farragut; for, though but a boy of twelve, he was selected to command the party of seamen detailed to manage the Barclay during this long passage. The captain of the Barclay went with his ship, but in great discontent that the command of the seamen was given not to himself, but to such a lad from the ship-of-war. Being a violent-tempered old man, he attempted by bluster to overawe the boy into surrendering his authority. "When the day arrived for our separation from the squadron," writes Farragut in his journal, "the captain was furious, and very plainly intimated to me that I would 'find myself off New Zealand in the morning,' to which I most decidedly demurred. We were lying still, while the other ships were fast disappearing from view, the commodore going north and the Essex Junior, with her convoy, steering to the south for Valparaiso. I considered that my day of trial had arrived (for I was a little afraid of the old fellow, as every one else was). But the time had come for me at least to play the man; so I mustered up courage and informed the captain that I desired the maintopsail filled away, in order that we might close up with the Essex Junior. He replied that he would shoot any man who dared to touch a rope without his orders; he 'would go his own course, and had no idea of trusting himself with a d—d nutshell'; and then he went below for his pistols. I called my right-hand man of the crew and told him my situation. I also informed him that I wanted the maintopsail filled. He answered with a clear 'Ay, ay, sir!' in a manner which was not to be misunderstood, and my confidence was perfectly restored. From that moment I became master of the vessel, and immediately gave all necessary orders for making sail, notifying the captain not to come on deck with his pistols unless he wished to go overboard, for I would really have had very little trouble in having such an order obeyed. I made my report to Captain Downes (of the Essex Junior), on rejoining him; and the captain also told his story, in which he endeavored to pursuade Downes that he only tried to frighten me. I replied by requesting Captain Downes to ask him how he succeeded; and to show him that I did not fear him, I offered to go back and proceed with him to Valparaiso. He was informed that I was in command, he being simply my adviser in navigating the vessel in case of separation. So, this being settled and understood, I returned to the Barclay, and everything went on amicably up to our arrival in Valparaiso."

It was on the 30th of June that the little squadron sailed from Tumbez, standing to the westward till they should reach the trade-winds; and on the 4th of July that the Essex Junior separated, with the prizes, and Farragut had his scene with the captain of the Barclay. As the winds on the west coast of South America blow throughout the year from the southward, the passage of sailing vessels in that direction is always long; but for the same reason the return is quickly made. When, therefore, the Essex Junior rejoined the Essex at the Galapagos, on the 30th of September, she brought comparatively recent news, and that of a very important character. Letters from the American consul in Buenos Ayres informed Porter that on the 5th of July the British frigate Phœbe, of thirty-six guns, a vessel in every way of superior force to the Essex, had sailed from Rio Janeiro for the Pacific, accompanied by two sloops-of-war, the Cherub and Raccoon, of twenty-four guns each. This little squadron was charged with the double mission of checking the ravages of the Essex and of destroying the fur trade of American citizens at the mouth of the Columbia River. From the date of their leaving Rio these ships were not improbably now on the coast; and allowing for time to refit after the stormy passage round the Horn, they might be expected soon to seek Porter at the Galapagos, the headquarters of the British whalers.

The Essex Junior brought back the prize-crews

and prize-masters who had navigated the captured ships to Valparaiso, and with the others Farragut now rejoined the frigate. During their absence Porter had taken four more valuable vessels. According to his information, there remained but one uncaptured of the British whalers which centered around the islands. The Essex had taken eleven; and among these, six carried letters of marque from their Government, authorizing them to seize for their own profit vessels of a nation at war with Great Britain. These powers would doubtless have been exercised at the expense of the unprepared American whalers but for the opportune appearance of the Essex, which had also released the vessels of her country from the ports to which, at the time of her arrival, they had been driven by Peruvian privateers. Porter's work in this region was therefore finished. He had entirely broken up an important branch of British commerce, inflicting damage estimated at nearly three million dollars; but the coming of an enemy's force considerably superior to his own, an event wholly beyond his control, reversed all the conditions and imposed upon him some new line of action. For this he was already prepared, and he took his decision with the promptitude characteristic of the man. The commander of the British squadron, Captain Hillyar, was personally well known to him, being an old acquaintance in the Mediterranean; and he doubtless realized from observation, as well as from his past record, that his enemy was not a man to throw away, through any carelessness or false feeling of chivalry, a single advantage conferred by his superior force. On the other hand, Porter himself was not one quietly to submit to superiority without an effort to regain the control which the chances of naval war might yet throw into his hands. He was determined to fight, if any fair chance offered; but to do so it was necessary to put his ship in the highest state of efficiency, which could only be done by leaving the spot where he was known to be, and, throwing the enemy off his scent, repairing to one where the necessary work could be performed in security. Two days after the arrival of the Essex Junior all the vessels sailed from the Galapagos Islands for the Marquesas. On the 25th of October they anchored at one of this group, called Nukahiva Island.

During the six weeks the Essex lay at this anchorage her crew bore a part in several expeditions on shore, designed to protect the natives in the neighborhood against hostile tribes in other parts of the island. In this land fighting Farragut and his younger messmates were not allowed to share; but were, on the contrary, compelled to attend a school established on board of one of the prizes, with the ship's chaplain for school-master. They were, however, permitted out of school hours and after the day's work, which for the ship's company ended at 4 P. M., to ramble freely in the island among the natives; considerable liberty being allowed to all hands, who, during their year's absence from the United States, had had little opportunity to visit any inhabited places. Farragut here learned to swim, and the aptitude of the natives to the water seems to have impressed him more than their other peculiarities which have since then been so liberally described in books of travel. "It appears as natural," he wrote, "for these islanders to swim as to eat. I have often seen mothers take their little children, apparently not more than two years old, down to the sea on their backs, walk deliberately into deep water, and leave them to paddle for themselves. To my astonishment, the little creatures could swim like young ducks."

On the 9th of December, 1813, the Essex and Essex Junior sailed for Valparaiso with one of the prizes, leaving the others at the Marquesas. Nothing of interest occurred during the passage, but the crew were daily exercised at all the arms carried by the ship—with the cannon, the muskets, and the single-sticks. The latter are for training in the use of the broadsword or cutlass, the play with which would be too dangerous for ordinary drills. Porter had a strong disposition to resort to boarding and hand-to-hand fighting, believing that the very surprise of an attack by the weaker party would go far to compensate for the inequality of numbers. On more than one occasion already, in the presence of superior force, he had contemplated resorting to this desperate game; and to a ship the character of whose battery necessitated a close approach to the enemy, the power to throw on board, at a moment's notice, a body of thoroughly drilled and equipped swordsmen was unquestionably of the first importance. "I have never since been in a ship," said Farragut at a later day, "where the crew of the old Essex was represented, but that I found them to be the best swordsmen on board. They had been so thoroughly trained as boarders that every man was prepared for such an emergency, with his cutlass as sharp as a razor, a dirk made from a file by the ship's armorer, and a pistol." With a ship well refitted and with a crew thus perfectly drilled, Porter had done all that in him lay in the way of preparation for victory. If he did not win, he would at least deserve to do so. For Farragut it is interesting to notice that, in his tender youth and most impressible years, he had before him, both in his captain and in his ship, most admirable models. The former daring to recklessness, yet leaving nothing to chance; fearless of responsibility, but ever sagacious in its exercise; a rigid disciplinarian, who yet tempered rigor by a profound knowledge of and sympathy with the peculiarities of the men who were under him. The latter—the ship—became, as ships under strong captains tend to become, the embodiment of the commander's spirit. Thoroughly prepared and armed at all points, she was now advancing at the close of her career to an audacious encounter with a greatly superior force. Whether the enterprise was justifiable or not, at least nothing that care could do to insure success was left to chance or to favor. Porter might perhaps have quitted the Pacific in December, 1813, and, reaching the United States coast in the winter, have escaped the blockade which at that season was necessarily relaxed. By doing so he might have saved his ship; but the United States Navy would have lost one of the most brilliant pages in its history, and its future admiral one of the most glorious episodes in his own great career.

On the 12th of January, 1814, the Essex arrived off the coast of Chile, making the land well to the southward—that is, to windward—of Valparaiso. From this point of arrival she ran slowly to the northward, looking into the old town of Concepion, between two and three hundred miles from Valparaiso. In the latter port she anchored on the 3d of February. The ordinary salutes and civilities with the authorities having been exchanged, every effort was made to get the ship ready for sea, the Essex Junior being employed cruising off the port so as to give timely notice of the approach of an enemy; a precaution necessary at all times, even in a neutral port, but especially so at a period when neutral rights were being openly disregarded in every direction by both the great belligerents, France and Great Britain. Moreover, Captain Hillyar, though a brave and experienced officer, a favorite with Nelson, whose esteem could not be won without high professional merit, was reputed to have shown scanty scruples about neutral rights on a previous occasion, when the disregard of them procured an advantage to the enterprise he had in hand. Being sent with several armed boats to attack two Spanish corvettes lying in the port of Barcelona, in the year 1800, he had pulled alongside a neutral vessel, a Swede, which was standing into the harbor; and after examining her papers in the due exercise of his right as a belligerent, his boats hooked on to her, thus using a neutral to tow them into the enemy's port, so that his men reached their scene of exertion unfatigued by the oar, and for a great part of the way protected by such respect as the Spanish batteries might show to a neutral coerced into aiding a hostile undertaking. "Having approached within about three quarters of a mile of the nearest battery," says the British naval historian James, "and being reminded by two shots which passed over the galliot that it was time to retire from the shelter of a neutral vessel, Captain Hillyar pulled away." Both the Spanish and Swedish Governments complained of this act, and their complaints delayed the promotion which Hillyar's gallantry would otherwise have won. Whatever the strict propriety of his conduct in this case, it was sufficiently doubtful to excite a just suspicion that Hillyar would not be deterred, by over-delicacy about the neutrality of the port, from seizing any advantage offered him by the unwariness of his enemy; and so the event proved.

On the 7th of February a dance was given on board the Essex, which lasted till midnight. In order that her officers might share in the entertainment, the Essex Junior was allowed to anchor, though in a position to have a clear view of the sea; but, when the guests began to depart, her commander went on board and got under way to resume his station outside. Before the decorations of the ball-room had been taken down, a signal was made from her that two enemy's ships were in sight. A whole watch—one third of the Essex's crew—were then on shore, but were quickly recalled by a gun. The ship was at once cleared for action, and the men at their quarters, with all the rapidity to be expected from the careful drilling they had had during their long commission. Porter himself had gone to the lookout ship to reconnoitre the enemy. Upon his return he found the frigate all ready for battle, it being then just an hour and a half since the alarm was given. The Essex Junior was then anchored in a position to support the Essex should occasion arise.

The strangers were the Phœbe and the Cherub. The third British ship, the Raccoon, had gone north to the Columbia. As has before been said, Captain Hillyar was an old friend of Porter's. The two men had been thrown together in the Mediterranean, and the American had been a frequent visitor in the other's house at Gibraltar. On one occasion Hillyar's family had made a passage from Malta to Gibraltar in an American ship-of-war; for in those troubled times would-be voyagers had to avail themselves of such opportunities as offered, and the courtesy of a large armed ship was among the most favorable. It was natural, therefore, that, as the Phœbe stood into the harbor, Captain Hillyar should bring his ship, the wind allowing it, close to the Essex and hail the latter with a polite inquiry after Captain Porter's health; but it was going rather too far, under all the circumstances, not to be content with passing slowly under the Essex's stern, than which no more favorable position could be found for an exchange of civil words. Instead of so doing, the helm of the Phœbe was put down and the ship luffed up into the wind between the Essex and the Essex Junior, the latter lying now near the senior ship and on her starboard beam. Whethei Hillyar counted upon his own seamanship to extricate his ship from the awkward position in which he had placed her, or whether, as the Americans believed, he intended to attack if circumstances favored, he soon saw that he had exposed himself to extreme peril. As the Phœbe lost her way she naturally fell off from the wind, her bows being swept round toward the Essex, while her stern was presented to the Essex Junior. Both her enemies had their guns trained on her; she could use none of hers. At the same time, in the act of falling off, she approached the Essex; and her jib-boom, projecting far beyond her bows, swept over the forecastle of the latter. Porter, who had been watching the whole proceeding with great distrust, had summoned his boarders as soon as the Phœbe luffed. The Essex at the moment was in a state of as absolute preparation as is a musket at full cock trained on the mark, and with the marksman's eye ranging over the sights; every man at his post, every gun trained, matches burning, and boarders standing by. The position was one of extreme tension. The American captain had in his hand a chance such as in his most sanguine dreams he could scarcely have hoped. His guns, feeble at a distance, could tell with the greatest effect at such short range; and even if his enemy dropped an anchor, in the great depths of Valparaiso Bay he would not fetch up till far past the Essex. Until then he was for the moment helpless. Porter hailed that if the ships touched he should at once attack. Hillyar kept his presence of mind admirably at this critical juncture, replying in an indifferent manner that he had no intention of allowing the Phœbe to fall on board the Essex—an assurance that was well enough, and, coupled with his nonchalant manner, served the purpose of keeping Porter in doubt as to whether a breach of neutrality had been intended. But the British frigate was unquestionably in a position where a seaman should not have placed her unless he meant mischief. It is good luck, not good management, when a ship in the Phœbe's position does not foul one in that of the Essex. While this was passing, Farragut was witness to a circumstance which shows by what a feather's weight scales are sometimes turned. Of all the watch that had been on shore when the enemy appeared, he says, one only, a mere boy, returned under the influence of liquor. "When the Phœbe was close alongside, and all hands at quarters, the powder-boys stationed with slow matches ready to discharge the guns, the boarders, cutlass in hand, standing by to board in the smoke, as was our custom at close quarters, the intoxicated youth saw, or imagined that he saw, through the port, some one on the Phœbe grinning at him. 'My fine fellow, I'll stop your making faces,' he exclaimed, and was just about to fire his gun, when Lieutenant McKnight saw the movement and with a blow sprawled him on the deck. Had that gun been fired, I am convinced that the Phœbe would have been ours." She probably would, for the Essex could have got in three broadsides of her twenty thirty-two-pounder carronades before the enemy could effectively reply, a beginning which would have reversed the odds between the two ships. Farragut fully shared the belief of all his shipmates that an attack was intended, in consequence of the information given to Captain Hillyar, as he was entering, by the boat of an English merchant ship in the port, that half the crew of the Essex was on shore. As the Phœbe luffed through between the two Americans a turn of her helm would have landed her on the bows of the Essex, if the latter had been caught at disadvantage. Instead of this, she was found fully prepared. The Essex Junior was also on the spot, while the Cherub, having drifted half a mile to leeward, could not have taken any part till the action was decided. Under these conditions, although their force was inferior, the advantage was with the Americans, whose ships were anchored and cleared, while the Phœbe still had her canvas spread and the anchoring to do, which is a troublesome operation in water so deep as that of Valparaiso Bay. If men's motives can be judged by their acts, Captain Hillyar afforded Porter full justification for opening fire. He extricated himself from a false position with consummate coolness; but his adversary, when taken later at disadvantage, had reason to regret the generosity with which he allowed him the benefit of the doubt as to his intentions to respect the neutrality of the port. As it was, when the two ships were almost touching, the Englishman threw his sails to the mast, and, backing clear of the Essex, anchored finally some distance astern.

The two British ships remained in port for a few days, during which their captains called upon Captain Porter on shore, where he was then living in the house of a gentleman named Blanco; and an amicable intercourse also grew up between the officers and crews of the two parties. Hillyar, however, told Porter frankly that he should not throw away the advantage given by his superior force, for the event of a naval action was ever uncertain, liable to be decided by the accidental loss of an important spar or rope; whereas, by keeping his two ships together, he thought he could effectually blockade the Essex and prevent her renewing her depredations upon British commerce until the arrival of other ships of war which were on their way. From this wary attitude Porter in vain tried to force his antagonist by varied provocations; but, although the exchange of official insults, verging closely at times upon personal imputations, caused bitterness to take the place of the first friendly courtesies, Hillyar was too old an officer, and his reputation for courage too well known, to allow his hand to be thus forced.

After filling with provisions and refitting, the

British ships left the anchorage and cruised off the approach to it, thus preventing the retreat of the Essex to the ocean, unless she could succeed in passing and then outsailing them. Valparaiso Bay is not an enclosed harbor, but simply a recess in the coast, which, running generally north and south, here turns abruptly to the eastward for two or three miles and then trends north again, leaving thus a concave beach facing the north. Along this beach lies the city of Valparaiso, stretching back and up on the hillsides, which rise to a height of twelve or fifteen hundred feet behind it. The prevailing winds along this coast being from the southward throughout the year, this formation gives an anchorage sheltered from them; but during the winter months of the southern hemisphere, from May to October, there are occasional northerly gales which endanger shipping, more from the heavy sea that rolls in than from the violence of the wind. In ordinary weather, at the season when the Essex was thus blockaded, the harbor is quiet through the night until the forenoon, when the southerly wind prevailing outside works its way in to the anchorage and blows freshly till after sundown. At times it descends in furious gusts down the ravines which cleave the hillsides, covering the city with clouds of dust and whirling sand and pebbles painfully in the faces of those who walk the streets.

On the 28th of March, 1814, such a blast descended upon the Essex, whose captain had by that time come to despair of forcing Hillyar to single combat. As the frigate straightened out her cables under the force of the wind, one of them broke, and the anchor of the other lost its hold upon the bottom. The

Essex began to drift to sea, and it was apparent would by this accident be carried out of reach of the port. Porter therefore ordered the cable cut and made sail on the ship, intending now to escape. The British ships kept habitually close to the western point of the bay; so that in case of such an attempt by their enemy he would have to pass to leeward of them, giving them a fair wind to follow. As Porter stood out, however, he thought possible, by keeping close to the wind, to pass to windward, which, with the superior sailing qualities of the Essex, would force the Phœbe to separate from the Cherub, unless Hillyar supinely acquiesced in his escape—an inadmissible supposition. If successful, he might yet have the single action he desired, and under conditions which would enable him to choose his distance and so profit by the qualities of his carronades. The Essex therefore hugged the wind; but as she was thus passing the western point of the bay, under a press of sail, a violent squall came down from the highland above, bearing the vessel over on her side and carrying away the maintopmast, which fell into the sea, drowning several of the crew. The loss of so important a part of her sail power made escape to sea impossible, and the Essex tried to regain the port. The wind, however, was adverse to the attempt in her crippled condition, so that she was only able to reach the east side of the bay, where she anchored about three miles from the city, but within pistol-shot of the shore, before the enemy could overtake her. As the conventional neutral line extends three miles from the beach, the Essex was here clearly under the protection of Chilian neutrality. Hillyar himself, in his official report of the action, says she was "so near the shore as to preclude the possibility of passing ahead of her without risk to His Majesty's ships." He seems, however, to have satisfied his conscience by drawing a line between the neutrality of the port and the neutrality of the country. The Essex was, he implies, outside the former. "Not succeeding in gaining the limits of the port, she bore up and anchored near the shore, a few miles to leeward of it."* At all events, having his adversary at such serious disadvantage, he did not propose to imitate the weakness Porter had shown toward himself six weeks before.

The crucial feature in the approaching action was that the Essex was armed almost entirely with carronades, and her principal enemy with long guns. The carronade, now a wholly obsolete arm, was a short cannon, made extremely light in proportion to the weight of the ball thrown by it. The comparative lightness of metal in each piece allowed a greater number to be carried, but at the same time so weakened the gun as to compel the use of a small charge of powder, in consequence of which the ball moved slowly and had but short range. In compensation, within its range, it broke up the hull of an enemy's ship more completely than the smaller but swifter ball from a long gun of the same weight; for the same reason that a stone thrown by hand demolishes a pane of glass, while a pistol-bullet makes a small, clean hole. It was this smashing effect at close quarters which gave the carronade favor in the eyes of one generation of seamen; but by 1812 it was generally recognized that, unless a vessel was

* Marshall's Naval Biography, article Hillyar, vol. iv, p. 861.

able to choose her own position, the short range of carronades might leave her helpless, and, even when she had the greater speed, an enemy with long guns might cripple her as she approached. Porter had begged to change his carronades for long guns when he joined the Essex. The request was refused, and the ship in this action had forty thirty-two-pounder carronades and six long twelve-pounders. The Phœbe had twenty-six long eighteen-pounders, one long twelve, and one long nine, besides eight carronades. The Essex being crippled and at anchor, Captain Hillyar, faithful, and most properly, to his principle of surrendering no advantage, chose his position beyond effective carronade range. The battle was therefore fought between the six long twelves of the Essex and the broadside of the Phœbe, consisting of thirteen long eighteens, one twelve, and one nine. Taking no account of the Cherub, the disparity of force is sufficiently obvious.

Although, from the assurances Hillyar had made to him in conversation, Porter had hoped that the neutrality of the port might be regarded, the manner in which the enemy's vessels approached his new anchorage gave serious reason to fear an attack. The ship was again got ready for action, and a spring put on the cable to enable the guns to be turned on the enemy in any position he might take. The desperateness of the situation was, however, manifest to all. "I well remember," wrote Farragut at a later day, "the feelings of awe produced in me by the approach of the hostile ships; even to my young mind it was perceptible in the faces of those around me, as clearly as possible, that our case was hopeless. It was equally apparent that all were ready to die at their guns rather than surrender; and such I believe to have been the determination of the crew, almost to a man." A crippled ship, armed with carronades, was indeed in a hopeless plight. At six minutes before four in the afternoon the attack began. The Essex riding to an anchor with a southerly wind, the Cherub took position on her starboard bow, or southwest from her; the Phœbe north, under her stern. Both British ships began fighting under sail, not being yet ready to anchor. The spring on the Essex's cable being shot away, she was unable to turn her broadside as was wished; but the Americans ran out of the stern-ports three of their long guns, which were so well served as to cut away some of the most important of the Phœbe's ropes and sails, and Hillyar for a moment feared his ship would be drifted out of action. The Cherub also was forced to leave her first position and join the Phœbe. The latter's damages being repaired, she regained her ground and anchored; both she and her consort placing themselves on the starboard quarter of the Essex, a position on which the American guns, neither from the stern nor the broadside, could be brought to bear unless by the springs on the cables. These, unfortunately, were three times shot away as soon as they had been placed. The first lieutenant of the Phœbe, a frank and gallant young Englishman, whose manly bearing had greatly attracted the officers of the Essex, is said to have remarked to his captain that it was no better than murder to go on killing men from such a position of safety, and to have urged him to close and make a more equal fight of it. Hillyar, so the story goes, replied that his reputation was established, and that as his orders were peremptory to capture the Essex, he was determined to take no risks. He might have added—probably did—that it was open to the Americans to save their lives by surrendering. The same view of the situation now impelled Porter, finding himself unable to give blow for blow, to try and close with his wary enemy. Only one light sail was left to him in condition for setting—the flying-jib. With it, the cable having been cut, the head of the Essex was turned toward the enemy; and, fanned along by the other sails hanging loose from the yards, she slowly approached her foes till her carronades at last could reach. The wary Englishman then slipped his cable and stood away till again out of range, when he resumed the action, choosing always his own position, which he was well able to do from the comparatively manageable condition of his ship. Finding it impossible to get into action, Porter next attempted to run the Essex aground, where the crew could escape and the vessel be destroyed. She was headed for the beach and approached within musket-shot of it, when a flaw of wind from the land cruelly turned her away.

The engagement had lasted nearly two hours when this disappointment was encountered. As a last resort, Porter now ordered a hawser to be made fast to an anchor which was still left. This was let go in the hope that, the Essex being held by it where she was, the enemy might drift out of action and be unable to return when the wind fell with the approaching sunset. The hawser, however, parted, and with it the last hope of escape. Great numbers of the crew had already been killed and wounded by the relentless pounding the ship had received from her enemies, for whom, toward the end, the affair became little more than safe target practice, with a smooth sea. As yet no voice had been raised in favor of submission ; but now entreaty was made to Porter to spare the lives of the remnant that was left, by ceasing a resistance which had become not only hopeless but passive, and which, however prolonged, could end only in the surrender of the ship. The latter had already been on fire several times, and was now alarmingly so, the flames rushing up the hatchways and being reported to be near the magazine. Porter then gave permission for such of the crew as wished, to swim ashore ; the colors being still flying, they were not yet prisoners of war. He next called his officers together to inform him as to the condition of the ship in the different parts where they served, but one only of the lieutenants was able to answer the summons. After consultation with him, satisfied that nothing more remained to be done, the order was given at twenty minutes past six to lower the flag of the Essex, after an action which had lasted two hours and a half. She had gone into battle with two hundred and fifty-five men. Of these, fifty-eight were killed, sixty-six wounded, and thirty-one missing. The last item is unusually large for a naval action, and was probably due to the attempt to escape to shore by swimming.

Farragut lacked still three months of being thirteen years old when he passed through this tremendous ordeal of slaughter, the most prolonged and the bloodiest of his distinguished career. At his tender years and in his subordinate position there could be, of course, no demand upon the professional ability or the moral courage which grapples with responsibility, of which he gave such high proof in his later life. In the Essex fight his was but to do and dare, perhaps it may rather be said to do and bear; for no heavier strain can be laid upon the physical courage than is required by passive endurance of a deadly attack without the power of reply. In the celebrated charge of the Six Hundred at Balaklava the magnificent display of courage was at least aided by the opportunity allowed for vehement action; the extreme nervous tension excited by such deadly danger found an outlet in the mad impetus of the forward rush. Farragut has himself recorded a singular instance in the Essex fight, which illustrates the sufficiently well-known fact that in the excitement of approaching action the sense of danger is subdued, even in a man who has not the strong nerves that endure the passive expectation of death. "On one occasion Midshipman Isaacs came up to the captain and reported that a quarter-gunner named Roach had deserted his post. The only reply of the captain, addressed to me, was: 'Do your duty, sir!' I seized a pistol and went in pursuit of the fellow, but did not find him. It appeared subsequently that when the ship was reported to be on fire he had contrived to get into the only boat that could be kept afloat, and escaped, with six others, to the shore. The most remarkable part of this affair was that Roach had always been a leading man in the ship, and on the occasion previously mentioned, when the Phœbe seemed about to run into us in the harbor of Valparaiso and the boarders were called away, I distinctly remember this man standing in an exposed position on the cat-head, with sleeves rolled up and cutlass in hand, ready to board, his countenance expressing eagerness for the fight; which goes to prove that personal courage is a very peculiar virtue."

Of his own courage the boy, in this his first action, gave the most marked proof. He was constantly under the captain's eye, and conducted himself so entirely to the satisfaction of that gallant officer as to be mentioned particularly in the dispatches. "Midshipmen Isaacs, Farragut, and Ogden exerted themselves in the performance of their respective duties, and gave an earnest of their value to the service." "They are too young," Porter added, "to recommend for promotion"—a phrase which Farragut thought had an ill-effect on his career, but which certainly implied that his conduct merited a reward that his years did not justify. During the action he was employed in the most multifarious ways, realizing the saying that whatever is nobody else's business is a midshipman's business; or, to use his own quaint expression, "I was like 'Paddy in the catharpins'—a man on occasions. I performed the duties of captain's aid, quarter-gunner, powder-boy, and, in fact, did everything that was required of me. I shall never forget the horrid impression made upon me at the sight of the first man I had ever seen killed. He was a boatswain's mate and was fearfully mutilated. It staggered and sickened me at first; but they soon began to fall around me so fast that it all appeared like a dream and produced no effect upon my nerves. I can remember well, while I was standing near the captain just abaft of the mainmast, a shot came through the waterways and glanced upward, killing four men who were standing by the side of the gun, taking the last one in the head and scattering his brains over both of us. But this awful sight did not affect me half as much as the death of the first poor fellow. I neither thought of nor noticed anything but the working of the guns. . . . When my services were not required for other purposes, I generally assisted in working a gun; would run and bring powder from the boys and send them back for more, until the captain wanted me to carry a message; and this continued to employ me during the action."

Although included in the report of the slightly wounded, Farragut received no serious injury, but he was not without the narrow escapes which must have been undergone by all the survivors of so desperate an action. One has just been related; and he has himself recorded two other incidents which came near making an end of him. "An old quartermaster named Francis Bland was standing at the wheel when I saw a shot coming over the fore yard in such a direction that I thought it would strike him or me; so I told him to jump, at the same time pulling him toward me. At that instant the shot took off his right leg, and I afterward found that my coat-tail had been carried away. I helped the old fellow below, and inquired for him after the action, but he had died before he could be attended to." At another time "some gun-primers were wanted and I was sent after them. In going below, while I was on the ward-room ladder, the captain of the gun directly opposite the hatchway was struck full in the face by an eighteen-pound shot and fell back on me; we tumbled down the hatch together. I struck on my head, and, fortunately, he fell on my hips. I say fortunately, for, as he was a man of at least two hundred pounds' weight, I would have been crushed to death if he had fallen directly across my body. I lay for some moments stunned by the blow, but soon recovered consciousness enough to rush on deck. The captain, seeing me covered with blood, asked if I were wounded, to which I replied: 'I believe not, sir.' 'Then,' said he, 'where are the primers?' This brought me completely to my senses, and I ran below and carried the primers on deck. When I came up the second time I saw the captain fall, and in my turn ran up and asked if he were wounded. He answered me almost in the same words: 'I believe not, my son; but I felt a blow on the top of my head.' He must have been knocked down by the wind of a passing shot, as his hat was somewhat damaged." The bruises from this fall down the hatch were the only injuries Farragut received.

When the surrender was determined, Farragut, at the captain's order, dropped the signal book overboard, watching it as it sank in the water till out of sight; and then in company with another midshipman amused himself throwing overboard the pistols and other small arms, to keep them out of the enemy's hands. The following morning he went on board the Phœbe, where the mortification of defeat drew tears from his eyes; a state of dejection from which he was roused by seeing a pet pig belonging to the Essex in the custody of one of the Phœbe's midshipmen. Farragut at once set up a claim to the porker as being private property, and as such to be respected by all civilized nations. The claim was resisted by the new owner; but his messmates, always ready for a lark, insisted that so doubtful a question must be decided by trial of battle. A ring being formed,

Farragut, after a short contest, succeeded in thrashing his opponent and regaining the pig, and with it a certain amount of complacency in that one Briton at least had felt the pangs of defeat. His grief mastered him again soon afterward, when asked by Captain Hillyar to breakfast with himself and Captain Porter. Hillyar, seeing his discomfiture, spoke to him with great kindness, saying: "Never mind, my little fellow, it will be your turn next perhaps"; to which, says Farragut, "I replied I hoped so, and left the cabin to hide my emotion."

After the action Porter and Hillyar entered into an arrangement by which the Essex Junior was disarmed and allowed to proceed to the United States as a cartel, under the charge of Lieutenant Downes, who had commanded her while a United States cruiser. All the survivors of the Essex except two, whose wounds did not permit, embarked in her and sailed from Valparaiso on the 27th of April for the United States, arriving on the 7th of July in New York. On the 5th, off the coast of Long Island, she was stopped by a British ship-of-war, whose captain questioned the right of Hillyar to give her the passports she carried, and indicated an intention of detaining her. Porter construed this violation of the stipulation between himself and his captor as releasing him from his obligations, and escaped to shore with a boat's crew. After a detention of nearly twenty-four hours the vessel was allowed to proceed; but was again overhauled by another British frigate as she approached Sandy Hook. There could be no serious question of detaining a ship that had been given a safeguard, under such circumstances and with such deliberation, by so experienced an officer as Hillyar.

But it is instructive to Americans, who are accustomed to see in the war of 1812 only a brilliant series of naval victories, to note that within a few hours' sail of their principal port British cruisers were lying in perfect security, stopping whom they would.

The Essex, upon which Farragut made his maiden cruise, and whose interesting career ended in so sad a catastrophe, remained, of course, in the hands of the victors. The little frigate was patched up and taken to England, where she was bought into the British Navy, and was borne on its register until 1837, when she was sold. After that all trace of her history is lost.

The Essex Junior, being a prize to the Essex and allowed to pass under Hillyar's safeguard, was sold in New York for the benefit of the captors.

NOTE.—The spelling Chile (instead of Chili) used in this chapter is that adopted by the United States Board on Geographic Names, appointed by President Harrison, September 4, 1890, to settle a uniform usage for the Executive Departments of the Government.

CHAPTER III.

MIDSHIPMAN TO LIEUTENANT.

1814–1825.

In common with the other survivors of the Essex, Farragut landed in the United States as a paroled prisoner of war. Captain Porter took him at once to Chester and put him again to school, this time to an old gentleman named Neif, who had served in the guards of Napoleon. The method of instruction practiced by him seems to have been unsystematic and discursive; but Farragut, who was ever attentive to make the most of such opportunities as offered for self-improvement, derived profit here also, and said afterward that the time thus passed had been of service to him throughout his life. Until very lately there were residents of that neighborhood who could recall the young midshipman as he was at Neif's school; a lad short of stature and not very handsome in face, but who bore himself very erect because, as he often declared, he could not afford to lose a fraction of one of his scanty inches. There was, and still is, near the spot where he went to school a tavern called the Seven Stars, which has been a public house since the time of the Revolution, and which had sheltered Howe and Cornwallis as the British army advanced from the head of the Chesapeake toward Philadelphia, in 1777. Upon its porch Farragut spent much of his leisure time, and within its walls joined in the social gayeties of the neighboring families, who afterward recalled with pride and interest this association with the young sailor before whom lay such a brilliant but unforeseen future.

In November, 1814, Farragut was exchanged, and at once ordered to New York to join the brig Spark, which was intended to form one of a squadron of small vessels to cruise against British commerce under the command of Captain Porter. He was here for the first time separated from his guardian and thrown wholly upon his own force of character to guide his steps; and this beginning was made with a set of messmates with whom he was temporarily quartered on board the John Adams, among whom were several very wild young men. Farragut evidently felt the force of the temptation, for he speaks with warm thankfulness of the counter-influence of the first lieutenant, to which he attributed much of his deliverance from the dissipation by which he was surrounded. "When I have looked back with a feeling of horror to that period of my life," he wrote, "I have always remembered with gratitude Mr. Cocke's counsels and kind-hearted forbearance." It was indeed characteristic of the man that, while by no means insensible to the natural temptations of youth, he was ever more attracted to and influenced by the good than by the evil around him. During the following year, on his cruise to the Mediterranean, he was messmate with a midshipman named William Taylor, a young man of singularly fine character, which seems to have been the chief cause of the influence he exerted upon Farragut.

"He took me under his charge, counseled me kindly, and inspired me with sentiments of true manliness, which were the reverse of what I might have learned from the examples I saw in the steerage of the John Adams. Never having had any real love for dissipation, I easily got rid of the bad influences which had assailed me in that ship." He noted also that, of the twelve or thirteen midshipmen there associated with him, in less than two years all but one, his old messmate Ogden, of the Essex, had disappeared from the navy. The habit of strict attention to duty which he had contracted under the rule of the Essex also contributed, by keeping him occupied and attentive, to deter him from yielding to practices incompatible with its due discharge.

The conclusion of peace put an end to the proposed cruise of the Spark, and Farragut was next ordered, in March, 1815, to the Independence, a seventy-four-gun ship, or ship-of-the-line, as such were commonly called. She was the flag-ship of a numerous squadron, composed mostly of small vessels, destined to act against Algiers, with whom war had recently been declared. Upon arriving in the Mediterranean it was found that Commodore Decatur had already brought the Dey to terms, so that Farragut saw here no more fighting, and the squadron returned home by winter. The following spring he was ordered to the Washington, also a seventy-four, about to sail for Naples, bearing on board Mr. William Pinkney, our minister to that court. This cruise gave our young midshipman an experience of a kind he had not before had, and which in more ways than one was useful to him. The Washington was one of those exceptional vessels which illustrated in the highest degree the kind and pitch of perfection to which, by unremitting severity and exaction, the appearance and drills of a ship-of-war could be brought. Her commander, Captain Creighton, had the reputation of being the greatest martinet in the navy; and being seconded by a singularly efficient and active set of officers, the ship was made to realize the extreme ideal of a naval officer of that day in smartness, order, and spotless cleanliness.* "But," says Farragut, "all this was accomplished at the sacrifice of the comfort of every one on board. My experience in the matter, instead of making me a proselyte to the doctrine of the old officers on this subject, determined me never to have 'a crack ship' if it was only to be attained by such means." His feeling on the matter was doubtless somewhat quickened by the personal discomfort which he, in common with all subordinates, underwent under such a system, although he was rather a favorite with the captain, whose aid he was; but it shows independence of character to have thought so clearly for himself at such an age, and to have ventured to differ from standards which were then, and for a long time afterward, implicitly accepted throughout the service. The tradition of those days, being mainly oral, has nearly disappeared; but fragments of it remain here

---

* The writer remembers to have heard in his early days in the service a tradition of a ship commanded by Creighton, which he believes to have been the Washington, and which illustrates the methods by which this extreme smartness was obtained. In each boat at the booms was constantly a midshipman in full dress, cocked hat included, so that no time might be lost in dropping alongside when called away. The full crew was probably also kept in her.

and there in the minds of those who, as youngsters thirty or forty years ago, were brought in contact with men, then already elderly, who had had personal experience of ships like the Washington. These stories, in their grotesque severities, have almost the air of an extravaganza. It must, however, be in justice remembered that they were the extravagances of a few among the men who had brought the United States Navy to the high efficiency in which it then was; and to whom, and not to either the people or the Government of that day, was due the glorious record of 1812. A few of them added to their military ardor and efficiency an undue amount of that spirit of the good housekeeper which makes a home unbearable. Farragut was aided to his wise conclusion by his previous experience in the Essex, where a high state of efficiency was gained without wanton sacrifice of comfort; for Porter, though a man of hasty temper, was ever considerate of his crew. But for the naval officers of that day Farragut throughout his life retained a profound admiration. Talking about them at his dinner-table in New Orleans fifty years later, but a few days before his famous passage of the Mobile forts, he said: "We have no better seamen in the service to-day than those gallant fellows Bainbridge, Decatur, Hull, Perry, Porter, and Charles Stewart; and," he added, "I must not forget to mention McDonough, and poor unlucky Lawrence, as splendid-looking a sailor as I ever saw. If I only had their chance and could lay the Hartford alongside of an English ship, I should like it better than fighting our own people." Some years later he again expressed the same feelings to the same friend, to whom the author is indebted for the communication of them. His own glorious career was then finished, and his life's work lay open to the mature reflection of his declining years, when he thus acknowledged his obligations to the heroes of his boyhood. "Isaac Hull," he said, "was as good a seaman as ever sailed a ship. If I have done the country any service afloat, it is in no small degree owing to the ambition and enthusiasm he created in me, when I was a youngster, by his fair fight with and capture of an English frigate. I always envied Hull that piece of good work." It is to be suspected that the Admiral always felt that something was lacking to the fullness of his cup, in that he had only been allowed to fight forts, and not ships like his own; and it is no small evidence of the generosity of his character that his enthusiasm was so aroused by the deeds of others. He spoke of the fight between the Kearsarge and the Alabama in as glowing terms as were aroused by his recollection of the Constitution and the Guerrière. "I had sooner have fought that fight," he wrote, "than any ever fought upon the ocean."

The Washington stopped a few days at Gibraltar, where the rest of the squadron were then at anchor; and then sailed with all of them in company to Naples. During the remainder of the year 1816 the ship cruised along the Barbary coast until the winter had fairly set in, when she with the other vessels repaired to Port Mahon. Although now so close to the spot where his race originated, Farragut's journal betrays no interest in the fact. He was still too young for the sentimental considerations to weigh much in his mind; and it was not till many years later, in the height of his glory as a naval commander, that he visited his father's birthplace, Ciudadela, the capital city of Minorca. In the following spring the squadron resumed its cruising and made quite a round of the Mediterranean west of Italy; the journal mentioning visits to Gibraltar, Malaga, Leghorn, Naples, Sicily, and the cities on the Barbary coast. Farragut made full and intelligent use of the opportunities thus afforded him for seeing the world; and his assiduous habit of observation did much to store his mind with information, which the circumstances of his early life had prevented his gaining in the ordinary ways of school and reading. He was fortunate also at this time in having the society of an intelligent and cultivated man, the chaplain of the Washington, Mr. Charles Folsom. The chaplain in those days was commonly the only schoolmaster the midshipmen had; and their opportunities of learning from him depended very much upon the pressure exercised by the captain to compel the attention of a set of boys. Mr. Folsom, however, was drawn to Farragut by the eager willingness of the latter to acquire, and by his sense of his deficiencies. The manly character which had resisted the temptations to low dissipation, and sought naturally the companionship of the better rather than the worse among his associates, also attracted him. The friendship thus formed became, through a series of incidents, the cause of an unusual opportunity for improvement being offered to Farragut. In the autumn of 1817 Mr. Folsom received the appointment of consul to Tunis, which had just been vacated. The summer cruising of the squadron was drawing to an end, and the winter quarters at Port Mahon about to be resumed. Therefore, while the Washington was lying in Gibraltar, Mr. Folsom wrote to the commander-in-chief, Commodore Chauncey, asking permission to take the young midshipman to spend the winter with him in Tunis, to pursue his education under his care. In the letter he spoke very earnestly of his pupil's zeal for improvement, of his close attention, and ready response to any effort on the part of his instructor. The letter is interesting also in its recognition of Farragut's still existing relations to Captain Porter, "to whose wishes this request can not be repugnant." The letter was dated October 14, 1817; and, the required permission being given, the two friends in the following month sailed from Gibraltar for Marseille as passengers in the sloop-of-war Erie. At Marseille a slight incident occurred which, while not quite creditable to our hero, may have interest as showing natural character. Spending the evening at the house of a Mr. Fitch, he was, much against his will, obliged to play whist, for which he had no fondness. "Not getting along very well with my hand, the party showed great impatience, and I thought were rather insulting in their remarks. One individual went so far as to dash his cards on the table in derision of my play, when I returned the compliment by throwing them at his head. I apologized to Mr. Fitch and retired, much mortified, but my temper had been sorely tried." The display of temper was scarcely more than the provocation justified; and it is noteworthy that during a period when dueling was so common Farragut, though quick to resent, appears never to have been involved in a serious personal difficulty.

Early in 1818 the Erie, carrying Mr. Folsom and

his pupil, arrived in Tunis, where the latter remained for nine months, pursuing his studies on the site of the ancient maritime empire of Carthage. He mentions particularly the subjects of mathematics, English literature, French, and Italian. For languages he had great natural aptitude, and in later life was able to converse in several. The monotony of study was varied by the society of the few but agreeable foreign families residing in Tunis, and by occasional excursions in the neighborhood; when the interest of the present was happily blended, under the guidance of such a man as Mr. Folsom, with thoughts upon the past grandeur and history of the Carthaginian empire and the Roman province which had successively flourished on that soil. In one of these excursions Farragut received a partial stroke of the sun, from the effects of which he suffered for many years.

The period of his stay in Tunis exceeded the original intention, but doubtless with the approval of the commodore. It was brought to a close in the fall of 1818 by an outbreak of the plague, which increased to such an alarming extent that Mr. Folsom felt compelled to send his charge away just when the approach of another winter of comparative idleness for the squadron would have justified a longer stay. But deaths in Tunis had risen to a hundred a day, and all the families were living in a state of complete isolation, the houses being barricaded against outsiders; therefore on the 9th of October Farragut departed in a Genoese brig for Leghorn. Thence, after a quarantine of forty days, he went to Pisa; and from there to Messina, where the squadron had assembled for the winter of 1818–'19.

The friendship between Farragut and Mr. Folsom did not end with this separation. The latter survived to the end of the civil war, and was thus privileged to follow the successful and great career of the admiral to whom, while yet an unformed boy, he had thoughtfully extended a helping hand. As late as 1865 letters passed between the two, showing that both cherished warm recollections of that early association; Mr. Folsom dating his, as though careful to make the coincidence, on the anniversary of the day when he parted with his pupil in the harbor of Tunis and returned alone to the plague-stricken city.

The officers of the United States squadron passed a gay winter in Messina in 1819. Farragut was not yet eighteen years of age, but his bodily development had kept pace with his mental, and he writes that he always held his own at this time in all athletic exercises. The succeeding spring and summer were again spent in routine cruising on board the Franklin, seventy-four, which had taken the place of the Washington. In the fall of 1819 the squadron was in Gibraltar; and there, "after much opposition," Farragut was appointed an acting lieutenant on board the brig Shark. This promotion, coming at so early an age, he afterward looked upon as one of the most important events of his life. "It caused me to feel that I was now associated with men, on an equality, and must act with more circumspection. When I became first lieutenant, my duties were still more important, for in truth I was really commander of the vessel, and yet I was not responsible (as such) —an anomalous position which has spoiled some of our best officers. I consider it a great advantage to obtain command young, having observed, as a general rule, that persons who come into authority late in life shrink from responsibility, and often break down under its weight." This last sentence, coming from a man of such extensive observation, and who bore in his day the responsibility of such weighty decisions, deserves most serious consideration now, when command rank is reached so very late in the United States Navy.

After a short year in the Shark Farragut was ordered to return to the United States, to pass the examination required of all midshipmen before they could be confirmed to the rank of lieutenant. No opportunity offering for passage in a ship-of-war, he embarked in a merchant vessel called the America. On the passage he found himself, with the ship, confronted by an apparent danger, which occasioned a display of the fearlessness and energy always latent in his character. Those were days when piracy was rife upon the seas in the neighborhood of the West Indies and of the Spanish Main. The system was an outgrowth of the privateering carried on by French and Spanish marauders, for they were little better, against both British and neutral commerce during the wars of the French Revolution and Empire; and it had received a fresh impulse from the quarrel then existing between Spain and her American colonies, which since 1810 had been in revolt against the mother country. Privateering, having booty as its sole motive, rapidly tends to indiscriminate robbery, if not held strictly responsible by the country using it; and the remote, extensive, and secluded shores of Cuba, Haïti, and the South American coast defied the careless supervision of the weak Spanish Government. When within a few days' sail of the United States, the America fell in with an armed brig showing the colors of the new Colombian republic; but a flag was little guarantee for the character of a vessel if other signs told against her. Farragut describes both captain and crew of the America as being so overwhelmed with fear that, though expecting no mercy, they entertained no idea of resistance. Under the circumstances he took command; and having, fortunately, as passengers two seamen from the squadron going home sick, these formed a nucleus around which rallied the courage of the others, paralyzed only through disuse. It was, however, the firmness of the lad of eighteen, supported by his position as an officer and acting upon the two men prepared to recognize him as such, that redeemed the others from imbecility to manhood. The incident had no results, the stranger proving to be a regularly commissioned cruiser, and treating them with civility. Farragut's thoughtful, not to say philosophical, turn of mind was shown in his recorded reflections upon the difference between the conduct of the man-of-war's men and the merchant seamen, which he justly attributed not to inherent difference of natural courage, but to the habit of arms and of contemplating danger under a particular form.

On the 20th of November, 1820, Farragut again landed in the United States, having been absent four years and a half. He felt himself a stranger, having left as a mere boy, and knowing no one but Commodore Porter and his family. His examination soon followed, and was passed; but apparently not quite to his own satisfaction. A period of comparative quiet followed, spent principally in Norfolk, Virginia, during which he formed the attachment which resulted in his first marriage. In May, 1822, he was again ordered to sea in the sloop-of-war John Adams, in which he made a short cruise in the Gulf of Mexico and to Vera Cruz, where the Spanish power in Mexico was then making its last stand in the well-known fortress, San Juan de Ulloa. The ship returned to the United States early in December, 1822, when Farragut found the Mosquito fleet, as it was called, fitting out against the pirates of the Caribbean Sea. Learning that it was to be commanded by his old captain, Commodore David Porter, he asked for and obtained orders to the Greyhound, one of the small vessels composing it, commanded by Lieutenant John Porter, a brother of the commodore.

Since the peace with Great Britain, Captain Porter had been a member of the Board of Navy Commissioners; a body of three officers appointed by an act of Congress passed early in 1815, whose duties were to administer the affairs of the navy under the supervision of the Secretary. Meanwhile the sufferings, not only of American property but of the persons of American citizens, from the prevalence of piracy in the Caribbean Sea, had become unendurable. Ordinary naval vessels were, from their size, unable to enforce a repression for which it was necessary to follow the freebooters and their petty craft into their lairs among the lagoons and creeks of the West India islands. The general outcry rousing the Government to the necessity of further exertion, Captain Porter offered his services to extirpate the nuisance; with the understanding that he was to have and fit out the kind of force he thought necessary for the service. He resigned his position on the board on the 31st of December, 1822; but before that date he had bought and begun to equip eight Chesapeake schooners, of fifty to sixty tons burden, of which the Greyhound, Farragut's new vessel, was one. He also built five rowing barges, unusually large, pulling twenty oars. With these, supported by the ordinary man-of-war schooners, of which several were already in the service, and by the sloops-of-war, he expected to drive the pirates not merely off the sea, but out of their hiding-places.

The commodore put to sea with all his squadron on the 14th of February, 1823. A northeast gale was at once encountered, but the tiny vessels ran through it without any harm. For the next six months Farragut was actively employed in the operations of the little fleet, the Greyhound being one of the five which were sent through the Mona Passage, between Porto Rico and Haïti, and thence ransacked the southern shores of the latter island and of Cuba as far as Cape San Antonio, where Cuba ends. There were many encounters between the pirates and the squadron, sometimes afloat, sometimes ashore, in several of which our officer served, forcing his way with his party through marsh and chaparral and cactus—a service often perilous, always painful and exhausting. His health fortunately held out through it; nor did he take the yellow fever, which, as the summer wore on, made sad havoc among both officers and men. Toward the end of his time he obtained the command of one of the Mosquito schooners, which, however, he held but for a short period; for, not having yet received his lieutenant's commission, he was relieved by the arrival of an officer of that rank. An interesting incident of this cruise was a meeting with his brother William, then already a lieutenant, whom he had not seen for thirteen years. Soon after that he obtained permission to visit New Orleans; and it is a curious coincidence that the vessel in which he took passage thither was carrying the first load of bricks to build Fort Jackson, one of the defenses of New Orleans, by the passage of which nearly forty years later he began his career as commander-in-chief. His father had then been many years dead; but he met his sister, with whom he had to make acquaintance after so long a separation.

The service of the Mosquito fleet was one of great exposure and privation. "I never owned a bed during my two years and a half in the West Indies," wrote Farragut, "but lay down to rest wherever I found the most comfortable berth." It was, however, effectual, both directly and indirectly, to the suppression of piracy; seconded as it was by the navy of Great Britain, interested like our own country in the security of commerce. Driven off the water, with their lurking-places invaded, their plunder seized, their vessels burned, their occupation afloat gone, the marauders organized themselves into bandits, and turned their predatory practices against the towns and villages. This roused the Spanish governors from the indolent complacency with which they had watched robberies upon foreigners that brought profit rather than loss to their districts. When the evil was thus brought home, the troops were put in motion; and the pirates, beset on both sides, gradually but rapidly disappeared.

This Mosquito war had, however, one very sad

result in depriving the navy of the eminent services of Commodore Porter. In 1824 a gratuitous insult, accompanied by outrage, offered to one of his officers, led him to land a party at the town of Foxardo, in Porto Rico, and force an apology from the guilty officials. Although no complaint seems to have been made by Spain, the United States Government took exception to his action and brought him to trial by court-martial. Porter confidently expected an acquittal, having proof that the outrage was wanton, and that the officials had engaged in it to protect some piratical plunder which had been taken into the place. He argued also that the wording of his orders from the department authorized his action. The court, however, found him guilty of an offense which was charged as "disobedience of orders, and conduct unbecoming an officer," and sentenced him to six months' suspension. The sentence was accompanied by the expression that the court "ascribes the conduct of the accused which is deemed censurable to an anxious disposition, on his part, to maintain the honor and advance the interest of the nation and of the service." Indignant at the result, Porter resigned from the navy and took service with the Mexican Republic. After spending there four years of harassing disappointments, the election of General Jackson to the presidency gave him a friend in power. He returned to the United States in October, 1829, under the encouragement of letters from persons closely connected with the new administration. The President offered to nominate him to his old position in the navy, but Porter declined "to associate with the men who sentenced me for upholding the honor of the flag." This, striking a kindred chord in Jackson's breast, elicited a warm note of approval, and he appointed the commodore Consul-General to Algiers. The conquest of that country by France put an end to the office before he could assume the duties. The President then nominated him to be Chargé d'Affaires to Turkey. He went there in August, 1831, became Minister Resident in 1839, and died in this post in 1843.

After his return from the Mosquito fleet, Farragut married, on the 24th of September, 1823, Miss Susan C. Marchant, the daughter of a gentleman of Norfolk, Virginia. He was at this time far from well; fever, which spared him while on that sickly service, having seized him upon arrival in a healthier climate. It was probably due in part to this that two years passed after his marriage before he again joined a ship. During this period he spent some weeks with his bride in the house of Commodore Porter, who had returned temporarily from his squadron to regain his strength after a severe attack of yellow fever. This was probably his last close personal association with his early benefactor, whom the issue of the trial afterward separated from his country; but the correspondence between the two continued through life, Farragut maintaining to the last a grateful recollection of kindness shown to him by one whom he termed his "most venerated friend and commander." As late as 1835, writing from Constantinople in reply to a letter received from his former ward, Porter, then an ailing and broken man, notices this trait in him: "I have found in yours that treasure of a grateful heart which should be so much prized. I have never looked for any other return than what my feelings gave me, and to find such sentiments of gratitude from you, after all others had forgotten that they had received any benefits from me, is truly refreshing to the feelings." The relations thus testified to are an honor to the memory of both.

CHAPTER IV.

LIEUTENANT.

1825–1841.

After the termination of his cruise in the Mosquito fleet, and up to the beginning of the Civil War, the story of Farragut's life is for the most part but the record of the routine service of a naval officer in times of peace—periods of distant foreign cruising succeeding to, and being again succeeded by, periods of employment on shore in some of the many duties connected with the administration of the navy. But while in their superficial aspect there is little to distinguish these monotonous years, with their occasional breaks of exceptional incident, from the ordinary experiences of all naval officers, the journal of Farragut shows an activity of mind, a constant habit of observation, especially in professional matters, and a painstaking diligence in embracing every passing opportunity for improvement, which reveal to some extent the causes of his subsequent great successes. It is not indeed always possible to trace the precise connection between this or that observation, this or that course of study, and the later results; it is rather in the constant habit of doing the best at every moment, and in the gradual formation of mental character and correct professional knowledge, that are to be found the fruits of the strenuous exertion made throughout his life by Admiral Farragut. It is a noteworthy, though by no means unprecedented, circumstance that these characteristics obtained little or no recognition during his early and middle career. Unlike the great British admiral, Nelson, no war occurred to bring his high qualities into notice; and, when lacking but a year of Nelson's age when he fell at Trafalgar, Farragut was vainly petitioning the Navy Department for the command of a sloop-of-war in the war with Mexico, although he alleged his intimate knowledge of the scene of operations, the close personal examination he had made of it, and the privilege he had had of witnessing an attack by a French squadron but a few years before.

The early age at which he had left his home, the long absences of his youth, and the death of his father, had all contributed to sever his associations with New Orleans; so that his marriage in Norfolk, as was the case with so many officers of his day, fixed that city as his place of residence when not at sea. It is worthy of remembrance, in connection with his firm determination at a later day to stand by the Union rather than by a section of the country, that the only home Farragut had known out of a ship-of-war was the Southern city where he had twice married, and where the general sentiment was contrary to the course he took. The interest of the fact lies not in its bearing upon the rights or wrongs of the great quarrel that all are now fain to forget, but in showing the rare strength of character which, sustained only by its own clear convictions, resisted the social and friendly influences that overcame so many others.

In August, 1825, Farragut was promoted to be lieutenant, and at the same time ordered to the frigate Brandywine, chosen to carry back to France Lafayette, who was just drawing to a close his memorable visit to the United States. The ship sailed from the capes of the Chesapeake in September, reaching Havre after a passage of twenty-five days. From there she went to England, and thence to the Mediterranean, returning to New York in May, 1826. After his arrival Farragut was detached and went to New Haven with his wife, who had become a great sufferer from neuralgia and continued to be an invalid during the remainder of their married life. While living in New Haven he availed himself of the opportunity to attend lectures at Yale College. After his wife's treatment was finished they returned to Norfolk, where he remained until October, 1828, attached to the receiving ship and living on board with Mrs. Farragut. Here the interest which he had showed in the improvement of his own mind was transferred to the ship's boys, most of whom did not even know their letters. Farragut organized a school for these waifs, who at that time were little accustomed to receive such care, and was gratified to find very tangible results in the improvement shown by them. He next received orders to the sloop-of-war Vandalia, which sailed from Philadelphia in the last days of 1828 for the Brazil station. On this cruise, which for him lasted but a year, he for the first time visited the Rio de la Plata and Buenos Ayres, and came in contact with the afterward celebrated dictator of that country, Rosas. The different provinces, whose union is now known by the political name of the Argentine Republic, had, under the later days of

Spanish rule, constituted with Bolivia, Paraguay and Uruguay the Viceroyalty of Buenos Ayres. On the 25th of May, 1810, a declaration of independence was issued in the city of Buenos Ayres. A long period of disturbance, internal and external, followed. At the time of this first visit of Farragut a contest had for some time been going on between two parties, representing two opposite political ideas, and striving in arms for the control of the State. The ideal of one was a strong centralized government supported by a powerful standing army. This naturally found its most numerous constituents among the wealthy and educated inhabitants of the principal city, Buenos Ayres. The province of the same name, however, and the other provinces generally, favored a looser form of confederation. The former party, known as the Unitarios, held a brief lease of power; but their opponents found an able leader in Juan Manuel de Rosas, who personified the best and worst features of the *gaucho* of the pampas and obtained unbounded popularity and following among those wild herdsmen. In 1828 Rosas and his allies forced the Unitarian president to resign, and installed one of themselves, named Dorrego, as governor of Buenos Ayres. This success was but one step in the series of bloody struggles which ended in the establishment of the dictator; but it marked the point at which Farragut first saw Buenos Ayres and Rosas himself, with whom he was at a later date thrown in intimate contact and who at that moment was in the full flush of his early popularity.

In December, 1829, Farragut's eyes were in such bad condition that it was found necessary to send him home. He arrived in February, 1830, and remained in Norfolk for a period of nearly three years, broken only by occasional absences. During a part of this time he was again attached to the receiving ship in the port; and, as before, manifested an interest, unusual in those days, in those under his command. One of these, then a midshipman, writes to the author that he still recalls, after the lapse of nearly sixty years, the kindness, consideration and hospitality shown him by the future admiral, who was then known through the service as the "Little Luff" Farragut—luff being a naval abbreviation, now obsolete, for lieutenant. But with all his kindness there was no relaxation in the enforcement of necessary duty. In December, 1832, he was again ordered to sea in the sloop-of-war Natchez, as her first lieutenant; or, as the expression now is, as executive officer. It was the time of the nullification troubles in South Carolina, and the ship was first sent to anchor near Charleston, where she would be prepared to support the authority of the United States Government. Fortunately, no occasion arose for her to act; and a stay which began with taking precautions against possible fire-ships from the city, ended in a series of balls and general exchanges of courtesy between the officers and the citizens. In April, 1833, the Natchez returned to Hampton Roads; and the following month sailed, carrying Farragut back again to the Brazils. On the 30th of July he was again at anchor, in his new ship, off Buenos Ayres. Since his former visit the country had passed through much trouble. A confederation had been formed between the principal provinces, in January, 1831, based upon the loosest ties of union; but the army had become dissatisfied with the progress of changes which arose largely from jealousy of the military power, and had risen in revolt under the leadership of a general named Lavalle, who for a time had sided with Rosas. He met at first with success, defeated Dorrego and Rosas, and put the former to death; but Rosas rallied again, defeated Lavalle, and became in his place head of the army and governor of Buenos Ayres. To this position he was re-elected in 1832, and by virtue of it he was, at the time of Farragut's second visit, in chief control of the external policy and internal affairs of the confederation; the principal and seaboard province inevitably taking the lead and representing the country under even the loosest form of combination. Disturbed though the internal state of affairs was, Rosas's strong hand appears to have so far preserved the safety of foreigners as to give no cause for the interference of their ships-of-war. Farragut's stay on the station was, however, again cut short. The schooner Boxer arrived in Rio Janeiro on her way home from the East Indies; and it becoming necessary to give her a new commanding officer, he received orders to take her to the United States. He sailed in her on the 8th of June, 1834, and on the 25th of July reached Norfolk, where the vessel was put out of commission and he again returned to his family. A period of nearly four years of shore duty followed. During the latter two of these Farragut was a constant applicant for sea service, which he could not obtain. His wife was at this time becoming ever weaker and weaker. "I was necessarily confined very much to the house," he writes, "for my wife was so helpless I was obliged to lift her and carry her about like a child." His tender and untiring devotion to the suffering invalid was no less conspicuous than his careful attention to the other duties of life, and was the constant remark of those who were witnesses of this sorrowful period.

In April, 1838, Farragut was again ordered to sea in the home squadron, and in the following August, though still only a lieutenant, took command, in Pensacola, of the sloop-of-war Erie; a position that could only be temporary, because belonging naturally to an officer of higher rank. It fell to him, however, at a period of peculiar interest—when France became involved with Mexico in one of those brief hostilities by which alone were broken the long years of peace between Waterloo and the Crimean War. The quarrel between the two was simply as to the reparation due to French subjects for injuries received during the long years of confusion through which Mexico then had been and still was passing. As a political question it possesses no present interest whatever; but to a naval officer of Farragut's strong professional feeling and close habits of observation it offered a peculiar opportunity for noting the silent progress made during the long peace by the material of war among the navies of Europe, where the necessity of constant preparation insures an advance in which the United States then, as now, tended to lag behind. It supplied also a test, under certain conditions, of the much-vexed question of the power of ships against forts; for the French squadron, though few in numbers, deliberately undertook to batter by horizontal fire, as well as to bombard, in the more correct sense of the word, with the vertical fire of mortars, the long renowned castle of San Juan de Ulloa, the chief defense of Vera Cruz. It was still the day of sailing-ships, both of war and of commerce. But a few years had elapsed since a man of considerable scientific attainment had pronounced the crossing of the Atlantic to be impossible to vessels depending upon steam power alone; and only in the same year as the French attack on Vera Cruz (1838) had been seen the falsification of the prediction by the passage of the Sirius and Great Western from England to New York.

As a first means of compulsion, the French Government had in 1837 established a blockade of the Atlantic ports of Mexico. In two months the Mexican treasury lost two million dollars in duties, which would have been collected if the ships turned away had been permitted to enter; but the Government and people seemed little moved by a result that merely added one more to the many ills with which they were already afflicted. The question was then raised by the French authorities, diplomatic and military, whether the possession of the fortress of San Juan de Ulloa, which commanded the city of Vera Cruz, the most important of the coast ports, would not also confer control of a great part of the seaboard, and thus enforce a security not otherwise obtainable for the persons and property of French subjects. Blockade, though a less extreme measure, was difficult, protracted, and productive of serious loss. The violent northerly gales of winter exposed the ships to peril, and the yellow fever of the summer months was deadly to the crews. Moreover, the deprivation of commerce, though a bitter evil to a settled community whose members were accustomed to the wealth, luxury, and quiet life attendant upon uninterrupted mercantile pursuits, had been proved ineffective when applied to a people to whom quiet and luxuries were the unrealized words of a dream. The French Government speedily determined to abandon the half-measure for one of more certain results; and in October, 1838, began to arrive the ships of an expedition destined to proceed to open hostilities, under the command of Admiral Baudin, a veteran of the Napoleonic wars. Appointed in the navy in 1799, immediately after the return from Egypt and the establishment of the Consulate, by the direct intervention of Bonaparte, who was a friend of his father's, Baudin had served with distinction until the fall of the empire, losing his right arm in battle; and after Waterloo it was he who made the proposition, familiar to all readers of Napoleon's life, to cover the escape of the Emperor from Rochefort by sacrificing the ships under his command in an heroic resistance to the English cruisers while the vessel bearing the fallen monarch escaped. "Sixteen years ago," said he, "my father died of joy upon learning the return from Egypt of General Bonaparte; and I myself to-day would die of grief to see the Emperor leave France if I thought that by remaining he could again do aught for her. But he must leave her only to live honored in a free country, not to die a prisoner to our rivals." Such was that career, belonging to an early and singular generation, which here for a moment crossed and linked with that of the great naval hero of our own days. Farragut has recorded his impression of him. "Admiral Baudin," he writes, "would be undoubtedly a *rara avis* in any navy. He is about fifty years of age (he was fifty-four), has lost his right arm, looks like a North of Europe man, has a fine address, and speaks English well. He has every mark of a polished seaman and officer, with the expression of great decision, with firmness and activity to execute his well-digested plans. These were my remarks the first time I saw him, and his subsequent conduct soon proved I was right." His French biographer makes a remark, commonplace enough, which yet notes the essential difference in the lot of the two gallant men who thus casually met. "For the few who allow occasions to escape them, how many could justly complain that a chance has never been offered them? Admiral Baudin never had the opportunity to which his capacities suited him; all his aptitudes designated him for war on a great scale; a man such as he, succeeding Latouche-Tréville, would have saved us the sorrows of Trafalgar." Farragut was fortunate, for in him the opportunity and the man met in happy combination.

When he reached his station, Admiral Baudin suffered no time to be lost. The wintry gales were approaching, while, on the other hand, his first experience showed the miseries of disease on that sickly coast. Of the two frigates there before he came, which had been blockading during the summer, one had buried forty-five seamen and five officers out of a ship's company of four hundred men; the other, at the time of his arrival, had three hundred and forty-three sick among a crew of five hundred. With such conditions, trifling is out of place. An ultimatum was at once sent to the Mexican Government, a brief time only being allowed for a reply, because the claims of the French cabinet were already clearly understood. On the 25th of November the last of his squadron, two bomb-vessels, arrived. On the 21st he had given notice that he would wait till noon of the

27th for the final decision. On the 28th the attack was made.

The castle of San Juan de Ulloa lies half a mile east and to seaward of the city of Vera Cruz, which it commands, and from which it is separated by water averaging from fifteen to twenty feet deep. It is built on the inner extremity of a reef that extends from it a little over a mile to the eastward, in the general prolongation of the line connecting the castle and the town. This shoal being covered by a foot or two of water, the builders of the fort counted upon it for protection in that direction against ships, and against attack either by regular approaches or by escalade. The work itself was in general outline a parallelogram, with bastions at the four angles. The longer sides fronted the east and west; and of these the former, facing the shoal and the open gulf, contained the gate of the fortress and was covered by a demi-lune and line of water batteries. There were mounted in the castle and dependent works, at the time of the French attacks, one hundred and eighty-six cannon. The strength of the fortifications, the number of the guns, and the character of the surroundings, had all contributed to bestow upon San Juan de Ulloa the reputation of being the strongest position in Spanish America. It was, indeed, considered impregnable to naval attack, for the best hope of ships under such circumstances is to approach closely and drive the defenders from their guns by the superior number and weight of the pieces opposed to them; but in San Juan this was forbidden by the extent and formation of the reef. Like most coral banks, this rises sheer from the bottom, making the approach very dangerous to vessels dependent only upon sail-power; and the ground about it, though not too deep for anchorage, is rocky and foul.

Admiral Baudin, however, was thoroughly acquainted with the weak points of the fortress, through information obtained from Madrid; where plans of the works, dating from the times of the Spanish occupancy, were on file. He possessed also two steamers, the first to cross the Atlantic under the French flag, by aid of which, though small and of weak power, he could count upon placing his sailing frigates exactly where he wished them. Finally, the wretched condition of the Mexican forces, demoralized by years of irregular warfare and internal commotion, and miserably provided with material of war, gave additional chances of success.

On the morning of November 28th the two steamers towed the bomb-vessels to the eastern extremity of the reef, a little over a mile from the castle. Next two of the frigates were taken by them and anchored close to the reef, southeast from the works and distant from them half a mile. The third frigate, using her sails alone, succeeded in taking position a little ahead of her consorts. These operations were all completed before noon and were conducted under the eyes of the Mexicans, who were restrained from impeding them by the orders of their Government not to fire the first gun. A delay followed, owing to a flag of truce coming from the shore; but the proposition brought by it proved unacceptable, and the squadron opened fire at half-past two. Between that and sundown the three frigates, aided only by a small corvette which attacked under way, poured upon the castle 7,771 round shot and 177 shell, the mortar-vessels at the same time throwing in 302 bombs. At eight the fire ceased, and negotiations began. The following day, at noon, the castle was delivered into the hands of the French, who placed a garrison in it. "It was high time," said Admiral Baudin; "the wind was freshening, the sea getting up, and the anchors were breaking like glass upon the bottom, composed of sharp rocks." But the loss among the defenders had been so great, and the re-enforcements at hand were so few, that further resistance was impracticable.

The terms of the convention made by the commander of the Mexican forces had stipulated that only a certain number of troops should constitute the garrison of Vera Cruz until the affairs between the two nations were settled; but upon the 4th of December the French admiral learned, to his great indignation, that the Mexican Government had disavowed the action of the general, declared war against France, and was throwing re-enforcements into the city. He immediately took measures to disarm the works which might threaten his fleet at their anchorage, hoping at the same time, by surprising the enemy, to gain possession of Santa Anna, the new commander of the troops and then the most prominent man in Mexico. While the French were making their preparations in secret, Farragut went on shore and called upon Santa Anna, who promised to care for the persons and property of American citizens, adding: "Tell President Van Buren that we are all one family, and must be united against Europeans obtaining a foothold on this continent."

The following morning, before daylight, the French embarked fifteen hundred seamen, accompanied by a few engineer soldiers, in the boats of the squadron; and, being covered by a thick fog, landed at six o'clock upon the beach before Vera Cruz. Formed in three divisions and unseen by the enemy, they blew open the gates of the city and at the same time stormed the forts which at the north and south terminate the seaward wall. The Mexicans, taken wholly by surprise, retreated before the assailants. The center division of the French, which had entered by the gates, pursued rapidly toward the quarters of Santa Anna. A short, vigorous resistance by a part of his guard enabled the commander-in-chief to escape in shirt and trousers; but General Arista was taken. Meanwhile the two flank divisions, having dismounted the guns in the forts and chopped the carriages in pieces, moved along the walls toward the gate. There they united with the center; and the whole body, having accomplished its object in disarming the sea face of the town, fell back upon their boats lying along the mole. Most had already re-embarked when the Mexicans, led by Santa Anna in person, charged from the gate and down the mole at double-quick. Admiral Baudin himself was still on shore, waiting to see the last man off. Though scarcely expecting this gallant return from a force that had been so badly worsted and was much inferior in numbers, the French were not unprepared. A six-pound gun on the extremity of the mole, belonging to the Mexicans, had been turned so as to sweep the approach with grape; and five of the boats of the squadron, mounting small carronades, were also disposed to repel attack. The admiral ordered the six-pounder fired, and entered his barge. The discharge swept away the head of the Mexican column, and Santa Anna himself fell with three wounds, from one of which he lost his left leg. Some of the broken column fell back upon the town, from the loop-holes of whose walls a sharp fire of musketry began, while others continued down the mole and opened vigorously upon the retreating French, directing their aim especially upon the admiral's barge. The admiral himself escaped, but narrowly; his cockswain and a midshipman standing by him being killed, and another midshipman wounded. "The Mexicans continued to fight with great gallantry," wrote Farragut; and it was perhaps well for the assailants that the fog sweeping in again covered their further retreat.

Of all these incidents Farragut was a close and interested observer. Upon joining the Erie as her commander, he found that the ship was under orders to proceed with the utmost dispatch to the Mexican coast, to afford to American citizens and their property the protection so likely to be needed in event of active hostilities. On the 26th of August she was anchored under the island of Sacrificios, off Vera Cruz, which was then still undergoing the blockade which preceded recourse to stronger measures. Farragut remained there till the 19th of September, when he returned to Pensacola; but early in November he was again off the Mexican coast at Tampico, where a revolution threatened, for Mexico at the time was not only menaced with foreign attack, but also a prey to the utmost internal disorder. On the 17th of this month the Erie ran down again to Vera Cruz; and learning there that the 27th was fixed as the day for a final conference and settlement of the questions at issue, her commander of course decided to remain throughout the affair, making preparations to receive on board Americans and their movable property in case the city was bombarded.

In his journal, and afterward in a letter to Commodore Barron, then the senior officer in the United States Navy, Farragut has preserved a very full and detailed account of the attack, the principal features of which have already been mentioned; and it is interesting to note, as testifying to the care and accuracy of his observations, that the account in his journal corresponds very closely with that given in the Life of Admiral Baudin, published in France within the last few years. He was particularly impressed with, and distinguishes as matters of principal importance, the utility of the small French steamers in towing the fighting ships into position, and the destructive effects of the shell upon the soft masonry of the fort. Admiral Baudin, in his reports, indulged in some of the pardonable grumbling of a seaman of the old school about the constant ailments of the little steam-vessels; but he was too capable an officer to ignore their value, "and never," wrote Farragut in his report, "was the utility of these vessels so apparent. Everything was done by them. The day was calm, or nearly so, and the ships had no sails to manage. As soon as the anchor was let go they were ready for action. The bomb-vessels were next placed (for which the range had been calculated), and two sloops took position at right angles with the range, to tell by signal the effect of the bombs. So you see all was arranged with science and skill and without the slightest interruption, for the Mexicans had given an order to the commander of the fort not on any account to fire the first gun." This order was, in Farragut's opinion, the principal cause of the French sustaining so little loss. A well-directed fire from the fort would, he thought, have destroyed the steamers and prevented the frigates from gaining the carefully chosen position, where they were least exposed to the guns of the works.

Immediately after the submission of the castle Farragut went ashore to examine and note the effects of the fire, and especially of the horizontal shell fire; which was then so much a novelty in naval warfare that he speaks of the missiles continuously as shell-shot, apparently to distinguish them from the vertically thrown bombs. "Now it was seen for the first time that the material of which Ulloa is built (soft coral) was the worst substance in the world for protection against the modern shell. The French threw almost entirely shell-shot, which entered the wall twelve or eighteen inches and then exploded, tearing out great masses of stone, and in some instances rending the wall from base to top. The damage done by these shell-shot was inconceivably greater than that by the shell from the bomb-vessels, owing to the former striking horizontally, while the latter fell vertically upon the bomb-proofs, doing but little damage. . . . I am satisfied of one fact—viz., that they might have bombarded with the bomb-vessels for a month without success, while the frigates would in four hours more, with their shell-shot, have reduced the fort to a heap of ruins." This opinion as to the inefficacy of bomb-firing to destroy a work anticipated the experience of the Civil War, where the conclusion was that it might wear out the endurance of the garrison by constant harassment, but not directly reduce the works themselves. It is only just to say that his estimate of the effect of the horizontal fire upon the walls is more favorable than that of the French engineers, who did not consider that the damage done necessarily entailed a capitulation; but seamen and engineers have rarely agreed in their opinions upon this subject.

The same zeal which led Farragut to this minute inspection of the battered fortress carried him also on board one of the French ships, while she still remained cleared for action, to note matters of detail which differed from those then prevalent in his own service. Of these he made a very full representation, and one much in disparagement of the United States Navy; which, since the glories of 1812 and the first re-organization and development procured for it by the popular favor consequent upon its victories, had been allowed to drop into a state of backwardness, as regards the material, similar to that which followed the Civil War, and from which it is but now beginning to emerge. The points which he noted, though most important to that rapidity and order upon which the efficient service of a ship's batteries depends, would have now no attraction for the unprofessional reader; nor for the professional, except as matters of antiquarian interest. They showed that spirit of system, of scientific calculation, of careful adaptation of means to ends, which have ever distinguished the French material for naval war, except when the embarrassments of the treasury have prevented the adoption of expensive improvements—a spirit which for over a century made the French ships the models which their usually victorious rivals were fain to copy. "The English and ourselves may affect to despise the French by sea," wrote Farragut to Barron, "but depend upon it, sir, they are in science far ahead of us both, and when England next meets France upon the ocean she will find a different enemy from that of the last war. Of all this I know you have seen much in theory, but I have seen it tested in practice."

The substance of Farragut's letter to Barron deals with matters which the progress of time and the accompanying advances in naval science have now made obsolete; but the spirit which inspired the letter and accumulated the materials for it can never become obsolete. It was then, and it is now, the indication of a man keeping abreast of his time and awake to its necessities; it held then, as it does now, the promise of one who, when occasion arose, would have his faculties in readiness, by constant training, to exert all the powers with which nature had gifted him. The conditions of 1861 were very different from those of 1838; but the officer who was found awake to the first in their day would not be behind the others in theirs. The letter concluded with a pregnant observation, which deserves to be quoted as thoroughly characteristic of the writer: "I have already said too much for a letter to any other person of your rank; but I flatter myself that I know your love of improvement, and that my intentions will be duly appreciated. If we who wander about the world do not keep those at home informed of the daily improvements in other navies, how can we hope to improve, particularly when we see men impressed with the idea that because they once gained a victory, they can do it again? So they may, but I can tell them it must be with the means of 1838, and not those of 1812." This transmission of information concerning the progress of other navies, upon which Farragut laid such just stress, is now systematized and perfected under a particular branch of the Navy Department, known as the Office of Naval Intelligence. Upon every ship afloat there is an officer whose duty is to observe and report to that office upon such matters, and upon all the experiences of foreign navies which are open to the examination of outsiders.

After the French affair at Vera Cruz the Erie returned to Pensacola, and there on the 12th of January, 1839, Farragut gave up the command to an officer of senior rank and went home. Upon his arrival in Norfolk, finding his wife's health to be very precarious, he remained unemployed until her death, which occurred on the 27th of December, 1840. "No more striking illustration of his gentleness of character," says his biography by his son, "is shown than in Farragut's attention to his invalid wife. His tenderness in contributing to her every comfort, and catering to every whim, through sixteen years of suffering, forms one of the brightest spots in the history of his domestic life. When not at sea, he was constantly by her side, and proved himself a faithful and skillful nurse. It was the subject of remark by all who were thrown with him; and a lady of Norfolk said, 'When Captain Farragut dies, he should have a monument reaching to the skies, made by every wife in the city contributing a stone.'"

CHAPTER V.

COMMANDER AND CAPTAIN.

1841–1860.

IMMEDIATELY after the death of his wife Farragut applied for sea service; and on the 22d of February, 1841, he was ordered to the Delaware, a ship-of-the-line, which was fitting for sea in Norfolk and destined to take him for the third time to the Brazil station. He was then among the senior lieutenants of the navy; but as it was in accordance with custom that a commander should be the executive officer of a ship-of-the-line, his expected promotion would not, when it arrived, cause him to leave his position. Some time passed before the Delaware was fully ready for sea. Before sailing, she was sent up the Chesapeake to the mouth of the Severn River, where she was visited by numbers of people from the neighboring city of Annapolis, as well as by large parties of congressmen and public officials from Washington, among whom came the then Secretary of the Navy. It was while lying off Annapolis, on the 27th of September, 1841, that Farragut received his commission as commander in the navy. His seniority as such was from September 8, 1841. A few days later the Delaware returned to Hampton Roads, and thence sailed for her station on the 1st of November. On the 12th of January she anchored in Rio Janeiro. After a stay of six weeks there, the whole squadron sailed for the Rio de la Plata, the usual resort of the ships on that station during the summer months of the southern hemisphere, when the yellow fever is apt to be prevalent in Rio Janeiro. On the 1st of June, 1842, Farragut was ordered to command the Decatur, a small sloop-of-war, relieving Commander Henry W. Ogden; who as a midshipman of the Essex had been his messmate nearly thirty years before, and was now compelled to leave his ship by an illness which never allowed him to resume the active pursuit of his profession. The transfer of the command appears to have been made in the harbor of Rio Janeiro. In severing his connection with the Delaware, with his new rank, Farragut felt that he had parted finally with the subordinate duties of his calling; and, as rarely happens, he passed directly from the active exercise of the lower position to fill the higher. His journal records the fact with a characteristic comment: "Thus closed my service on board the Delaware as executive officer; to which I shall always look back with gratification, as it was the last step in the ladder of subordinate duties, and I feel proud to think I performed it with the same zeal as the first." He was then nearly forty-one years old.

On the 2d of July the Decatur sailed for the La Plata in company with the Delaware. Soon after reaching Montevideo, Commodore Morris embarked on board the former, and went in her to Buenos Ayres; ships of the size of the Delaware not being able to approach that city on account of the great distance to which very shoal water extends from it. After exchanging the usual official civilities and transacting some business with Rosas, who then embodied in his own person all the powers of the state, the commodore returned to Montevideo; but the Decatur was soon sent back, and Farragut spent most of the latter half of 1842 at Buenos Ayres, in constant intercourse, both official and social, with Rosas and his family. Of the latter he, in common with most American naval officers who visited the La Plata at that time, received very agreeable impressions; and since, as commanding officer, his duties were less exacting and his time much more at his own command than as executive, he gave free play to the social disposition which was prominent in his character. Much of his journal during his stay is taken up with the accounts of social and official entertainments in which he shared. "During the month of September," he writes, "I made it a rule to spend two or three evenings a week at the governor's" (Rosas). "On the 5th of November I was invited to a ball at the Victoria Theatre, where, as on all similar occasions, I danced the first quadrille with the charming 'Manuelita," the daughter of Rosas. The pleasant and familiar relations thus established enabled him to do many kind acts for the Unitarios, whose lives were in constant danger by political accusations, if not from actual offenses.

Rosas himself was then in the full exercise of the dictatorial power with which he had been invested some years before, after refusing a re-election as governor of Buenos Ayres. His rule, which lasted under successive renewals of his office until 1852, was arbitrary and bloody; but in the disorganized condition of the provinces at that period a man of his force of character seems to have been necessary, to avert the greater horrors of constant intestine strife. "We concluded from our observations," notes Farragut in his journal, "that he was a man of uncommon mind and energy, and, as a general thing, reasonable; but on the subject of secret societies he was a madman, if we might judge from his furious denunciation of them." They constituted, indeed, the one resource of the cowed Unitarios, and were the chief danger then threatening him. "We had an excellent opportunity to form an idea of his character, as he appeared to throw off all restraint while with us. But the commodore informed us that, as soon as he laid business matters before him, Rosas was a different person; he was calm and measured in manner and language." The ladies of the family were amiable, intelligent and hospitable; but, like all the women of Buenos Ayres at that time, were perforce ardent Federalists and detesters of the "savage Unitarios." Farragut mentions an incident occurring at an official festivity in honor of Rosas, which shows the savagery that lay close under the surface of the Argentine character at that time, and easily found revolting expression in the constant civil strife and in the uncontrolled rule of the dictator. "In the ball-room was a picture which would have disgraced even barbarian society. It was a full-sized figure representing a Federal soldier, with a Unitarian lying on the ground, the Federal pressing his knees between the victim's shoulders, whose head was pulled back with the left hand, and the throat cut from ear to ear, while the executioner exultingly held aloft a bloody knife and seemed to be claiming the applause of the spectators. I am sure I do not err in saying that every one of our party felt an involuntary shudder come over him when his eye fell upon this tableau; nor did we afterward recover our spirits, everything in the way of gayety on our part during the night was forced and unnatural."

It is a matter of some, though minor, interest to note that Farragut has occasion at this time to mention Garibaldi, in connection with the wars then waging. The Italian patriot, whose name was then far from having the celebrity it has since attained, had for some time been engaged on the popular side in revolutionary struggles in the southern provinces of Brazil. Thence he had passed into Uruguay, and become a teacher of mathematics in Montevideo. Rosas had the ambition to bring into the Argentine confederation all the provinces which once formed the viceroyalty of Buenos Ayres, of which Uruguay was one; and, finding a pretext in the civil dissensions of the latter, had opened hostilities as the ally of one party in the State. Garibaldi, who began life as a seaman, had command of the Uruguayan naval forces, and in that capacity undertook to carry stores to Corrientes, an important point far up the river Parana. "As he met with many obstacles in his course," notes Farragut, "the Argentine admiral, Brown, was enabled to overtake him. Garibaldi ran his vessel into a creek and made a most desperate resistance; fought until he had expended everything in the way of ammunition, then landed his crew and set his vessel on fire." On the 17th of October a grand ball was given in honor of this success, which Commander Farragut attended; as he did all the other gayeties during his stay in Buenos Ayres.

The Decatur had already been long on the station when Farragut assumed command, and the time had now arrived for her to return home. After leaving Buenos Ayres she made short stops at Montevideo, Rio Janeiro, Maranham, and Para, the latter being the seaport of the Amazon River. On the 18th of February, 1843, she arrived in Norfolk, and Farragut was relieved. His health being delicate at this time, he spent the following summer at Fauquier Springs, Virginia.

From the mountains he returned in the autumn to Norfolk; and there on the 26th of December, 1843, he married Miss Virginia Loyall, the eldest daughter of Mr. William Loyall, a well-known and respected citizen of Norfolk.

In April, 1844, Commander Farragut was ordered as executive officer to the receiving ship at Norfolk, the Pennsylvania, of one hundred and twenty guns; which, in the days of sailing ships, was by far the largest vessel the United States ever had, and one of the largest in the world. Some time later he was transferred to the navy yard at the same place, on which duty he was employed when the war with Mexico arose.

As soon as the already existing difficulties with that country began to wear an ominous outlook, Farragut wrote to the Navy Department, asking for service in the Gulf. In his application he stated the qualifications he thought he possessed, from his knowledge and close study of the ground, and from his acquaintance with the Spanish language. He instanced particularly the occasions on which he had been employed in that neighborhood, and the close study he had been privileged to make on the spot during Admiral Baudin's operations. Although the Secretary of the Navy at that time was the able and enlightened Mr. George Bancroft, this letter received no reply; and a second, sent after the beginning of the war, was barely acknowledged without any action being taken. After Mr. Bancroft left the Department, Farragut renewed his application, expressing a decided opinion that the castle of San Juan de Ulloa could be taken either by artillery attack or by escalade; offering to undertake the task with the Pennsylvania and two sloops-of-war. If not thought to have rank enough for such a command, he was willing to go back to the position of executive officer of the Pennsylvania, in order, in that capacity, to organize the crew for the attack. The opinion thus expressed ran counter to the routine prejudices of the day, and, coming from an officer who had as yet had no opportunity to establish his particular claim to be heard, rather hurt than improved his chances for employment. It was not till February, 1847, nearly a year after the war began, and then with "much difficulty," that he obtained command of the sloop-of-war Saratoga; but when he reached Vera Cruz in her, the castle had already passed into the hands of the United States, having surrendered to the forces under General Scott on the 26th of March. That this capture should have been made by the army rather than by the navy was a severe disappointment to Farragut, who had so long cherished the hope that its fall should have been the brilliant achievement of his own service. In his mortification he used an expression which, in the light of his own subsequent career, seems a twofold prophecy. "The navy would stand on a different footing to-day if our ships had made the attack. It was all we could do, and should have been done at all hazards. Commodore Conner thought differently, however, and the old officers at home backed his opinion; but they all paid the penalty—*not one of them will wear an admiral's flag*, which they might have done if that castle had been taken by the navy, which must have been the result of an attack." It was to such enterprise at the hands of the men of his own time, among whom he was foremost, that the navy at a later day did obtain the admiral's flag which it had so long in vain desired.

The frustration of this high ambition was not the only misfortune to Farragut arising out of the Mexican war. He contracted the yellow fever on the station, nearly losing his life; and subsequently became involved in a controversy with the commodore of the squadron, who he believed had, in the assignment of duty, treated him and his ship with unfair discrimination, due to personal ill-will toward himself. The correspondence had no results; but such quarrels are rarely other than hurtful to the junior officer engaged. It is not singular, therefore, that he speaks of this cruise as the most mortifying of all the service he had seen since entering the navy. "I have little," he said again, "to look back to with satisfaction or pleasure at this time, except the consciousness of having done my duty." Smarting under the belief that he was being imposed upon, he wrote to the Navy Department complaining of injustice, and asking that either he himself should be relieved or the ship sent home. He candidly admits that his tetters were considered improper by the Secretary of the Navy, but the Saratoga was ordered to return to the United States, and was paid off at New York in February, 1848. In her short cruise there had been one hundred cases of yellow fever in her crew of one hundred and fifty, and her commander had been obliged, to use his own expression, "to rid the service" of five of her junior officers, and on the last day to bring the first lieutenant to trial for drunkenness. Altogether, the Mexican war and the cruise of the Saratoga seem to have marked the lowest point of disappointment and annoyance that Farragut was called upon to encounter during his naval career.

Immediately after leaving the Saratoga, Farragut was again ordered to duty in his former position at the Norfolk navy yard. Two years later he was called to Washington to draw up, in connection with some other officers, a book of Ordnance Regulations for the navy. This occupied him for eighteen months. As when in New Haven, twenty-five years before, he had improved the opportunity of hearing the lectures at Yale College, so at this later period he attended regularly those of the Smithsonian Institution, losing, he records, but a single one. "You will rarely come away from such lectures," he adds, "without being somewhat wiser than you went in." Where precisely such knowledge might come into play he could not, indeed, foresee, but he acted always on the principle that any knowledge might at some time become useful; just as, when at Vera Cruz, though he did not at the time look forward to a war with Mexico, he closely examined every point of interest, for "I have made it a rule of my life to note these things with a view to the possible future."

When the Ordnance Regulations were finished, in the spring of 1852, Farragut was again assigned to the Norfolk navy yard, and directed to utilize the experience he had gained in compiling them by giving weekly lectures on gunnery to the officers on the station. In prosecution of the same line of professional work, he was soon after ordered to conduct a series of experiments at Old Point Comfort, near Norfolk, to determine certain questions connected with the endurance of iron cannon; the discharges being continued with one or two of each class of service guns until they burst. Some very important results were obtained; but the circumstance connected with this duty which has now most interest, is that in it Farragut was associated with Lieutenant Percival Drayton, who was afterward his flag-captain and chief-of-staff at the battle of Mobile Bay. The intimacy formed during this year of experimental duty at Old Point lasted throughout their lives.

Soon after this the Crimean war broke out. Farragut's desire for his own professional improvement and for the progress of the service led him to make application to the Navy Department to be sent to the seat of war, "to visit the fleets of England and France, and ascertain whether in the outfits and preparation for war they possess any advantages over our own ships-of-war, and, if so, in what they consist." The utility of such a mission can not be doubted, and his occupations of the past few years particularly prepared him for such an inquiry. Had the Navy Department then had any systematic record of the aptitude shown by individual officers, and of the work done by them, it must have recognized Farragut's peculiar fitness for duties of this kind; which have since his time been organized and given a most comprehensive scope under the Intelligence Office of the Navy Department. As it was, his application received no other reply than a polite acknowledgment.

A commission, consisting of three officers of the Engineer Corps of the army, was sent by the War Department to visit Europe and the seat of war, and upon its return made an elaborate report; but at this critical period of naval progress, when sail was manifestly giving place to steam, when the early attempts at iron-clad batteries were being made, and the vast changes in armament that have since taken place were certainly, though as yet dimly, indicated, it did not appear to the Government of the United States a matter of sufficient importance to inquire, on the spot, into the practical working of the new instruments under the test of war.

Although doubtless not so intended, the Navy Department emphasized its decision not to send Farragut to the East by assigning him to duty as far west as the naval interests of the United States, within its own borders, then allowed. In August, 1854, four months after his application for the former employment, he was ordered to California as first commandant of the navy yard at Mare Island. The site had been selected in the year 1852 by a commission of three officers, but as yet no navy yard existed. It was to be Farragut's particular duty to plan and build it up under the general instructions of the Department. His selection for this difficult and onerous, but at the same time very flattering, appointment was among the first evident results of the diligent, painstaking effort which had marked his professional career. By that, and by that only, had he as yet had any opportunity of marking himself above the ordinary run of men; but he stood high in the esteem of Commodore Joseph Smith, then and for many years both before and after, the chief of the Bureau of Yards and Docks, under whose charge the management and development of navy yards more particularly came. At the critical period when the selection of an officer to command in the attack upon New Orleans had to be made, Smith, who had close confidential relations with the Secretary of the Navy, always held that Farragut was the man above all others for the place.

The site of the new yard was in the extensive sheet of inland waters connected with the bay of San Francisco, and some thirty miles from the city. There being no accommodations upon the island, Farragut, with his family, for some seven months lived on board an old sloop-of-war anchored near by. He remained at this station for four years, during which great progress was made in the development of the yard; but the duty, though most important and particularly responsible, because of the length of time required by correspondence to pass to and from Washington, was not fruitful of incident. These were the troublous early times of California—the days of the Vigilance Committee and the Law and Order Party. With these intestine troubles of a State the military officers of the United States had no proper concern; but there was continually a possibility that they might be forced to take a stand by the interference of one side or the other with civil officials of the United States Government, or might be induced, by a request from the authorities, to act upon the ground that there was no time to refer to Washington for instructions. It is unnecessary to enter into any examination of Farragut's course during this period, although the affairs with which he had to deal became at times both critical and delicate. It will be sufficient to say that the Navy Department, after receiving his reports, approved his conduct as having been prudent and yet marked by a proper spirit.

In July, 1858, Farragut returned to the East by the only route then available, the Isthmus of Panama. During his absence, on the 14th of September, 1855, he had been promoted to the rank of captain, which, prior to the Civil War, was the highest grade in the United States Navy; the title commodore, then so frequently applied to the older officers of the service, being simply one of courtesy given to a captain who had commanded a squadron of several vessels, but who did not thereby cease to be borne as a captain upon the Navy Register. Soon after his arrival Farragut was ordered to command the Brooklyn, one of six steam sloops-of-war just being completed. She belonged to that new navy of thirty years ago which the United States Government, most luckily for itself, had determined to build, and which became fairly available just in time for the exigencies of the Civil War.

It has been said, and that on the floors of Congress by a politician conspicuous in his party, that past history teaches that preparation for war is unnecessary to the United States, and the conditions precedent to the wars of 1812 and 1861 have been cited in support of the assertion. Certainly no one cognizant of the facts will deny that the United States was most miserably unprepared for either war as regards the size of her navy; but it so happened on both occasions, more by good luck than good management, that what navy it did have was of remarkably fine quality, and, to the extent to which its numbers permitted it to be employed, was generally perfectly adequate to the work it had to do. It could not, however, begin to touch the full amount of service it ought to have done. In 1812 it could not protect the Chesapeake nor the Mississippi; it was blockaded in its own ports, escaping only by evasion; it could not protect American commerce, which suffered more than did that of Great Britain. In 1861, had its numbers been at all adequate, it could by prompt action have forestalled the preparations of the enemy, and by prevention secured immediate advantages which were afterward achieved only by large expenditure of time and fighting. Such were the results of unpreparedness. It was to the preparation, scanty as it was—to the fine ships and superior armaments, both too few—that the successes of either era were due. The frigates and sloops of 1812 were among the finest of their class to be found anywhere, with powerful batteries and excellently officered; while in the decade before the Civil War began there had been built eighteen or twenty new steamships, admirably efficient for their day, and with armaments of an advanced and powerful type. Upon these fell the principal brunt of the naval fighting that ensued. These ships, and particularly those of the Brooklyn class, were the backbone of Farragut's fleet throughout all his actions, even in the last at Mobile in 1864. Had there been thrice as many, the work would have been sooner and therefore more cheaply done; but had the lack of preparation in 1861 equaled that of 1851 or 1881, it may be questioned whether any of his successes could have been won.

When Farragut took command of the Brooklyn, ten years had elapsed since he was last afloat—years pregnant with naval change. He had never before served in a steamer, except for a very short time in a primitive one belonging to Porter's Mosquito fleet, in 1823. The changes in the disposition and handling of the guns had not been radical. They were still arranged "in broadside," along the two sides of the vessel; nor were the pivot guns—which, as their name implies, could be pivoted to one side or the other, according to the position of an enemy—a new idea. In these matters there had been improvement and development, but not revolution. But while the mode of placing and handling was essentially the same, the guns themselves had greatly increased in size and received important modifications in pattern. The system then in vogue was that associated with the name of the late Admiral Dahlgren. The shape of the gun had been made to conform to the strains brought by the discharge upon its various parts, as determined by careful experiment; and in place of the 32-pounder, or six-inch gun, which had been the principal weapon of the earlier ships, the batteries of the new frigates and sloops were composed chiefly of nine-inch guns, with one or more pivots of ten- or eleven-inch bore. The shell-shot, whose destructive effects had excited Farragut's comments in 1838, were now the recognized type of projectile; and the new guns were spoken of distinctively as shell-guns, because not expected to use solid shot under ordinary circumstances. The Brooklyn and her fellows, among which was Farragut's future flag-ship, the Hartford, although screw steamers, had also the full sail power of the former sailing ship; and they were wooden, not iron vessels.

The service of the Brooklyn, while under Farragut's command, was chiefly confined to his old cruising ground in the West Indies and in Mexico. In the latter country, since the termination of the war with the United States in 1848, there had been a constant succession of revolutions; and at the time of the Brooklyn's cruise there was established in Vera Cruz a constitutional party, at whose head was Benito Juarez, the lawful claimant of the presidency. Opposed to this, in the city of Mexico, was the party headed by General Miramon, who had succeeded by force to the authority of Juarez's predecessor. The United States threw its influence on the side of Juarez; and its minister, Robert McLane, was permitted to use the Brooklyn to carry him from point to point of the coast. While no force was exerted, the support given to the minister's remonstrances by the constant presence of a powerful ship-of-war served to emphasize the policy of the Government, which had recognized Juarez. This recognition was followed some time later by a similar step on the part of the ministers of England, France, and Spain. Mr. McLane continued with the Brooklyn during great part of 1859, and in December of that year returned in her to the Mississippi, where he was landed at a plantation below New Orleans. This visit to his early home was marked by a sad coincidence to Farragut. His elder brother, William, a lieutenant in the navy, had long been retired from active service, for which he was unfitted by rheumatism. In consequence he had not received promotion, remaining at the head of the list of lieutenants, and being assigned to duty at the naval rendezvous in New Orleans. When the Brooklyn entered the river he was lying at the point of death, but heard of his brother's approach, and expressed a hope that he might live long enough to see him again after so many years of separation. The wish was not to be fulfilled. Though ignorant of the danger, Captain Farragut hastened to the city, himself also looking forward with pleasure to the meeting; but he arrived only in time to see his brother dead, and to follow him to the grave.

Farragut remained attached to the Brooklyn for two years. In October, 1860, he was relieved by Captain W. S. Walker, and returned to his home in Norfolk. This ended his sea service prior to the Civil War, and as the captain of a single ship. Thenceforward, during the brief but important remnant of his active career, he was to command great fleets.

CHAPTER VI.

THE QUESTION OF ALLEGIANCE.

1860–1861.

When Captain Farragut returned to Norfolk in October, 1860, he was, albeit unconsciously, rapidly approaching the turning point of his life, the tide in his affairs which taken at the flood should lead on to fortune. That he seized the opportunity was due to no dexterous weighing of the effects of either course upon his personal future, but to that preparedness of mind which has already been mentioned as one of his characteristic traits, and to the tenacity with which were held his convictions thus deliberately and maturely formed. For several years he had watched with unquiet mind the gathering clouds which preceded the approaching storm, and in common with others had felt the distress and perplexity which would attend the rupture of the Union. He did not, however, remain a merely passive spectator, agitated as such by hopes and fears, but trusting withal to the chapter of accidents. He had considered the effect of the alternatives before the country, and what his own duty should be in any case. He could not, in his modest position, control the course of events; but, whatever befell, he would be ready to take his stand, strengthened in so doing by the settled principles to which his conscientious meditation had led him. Thus his fixed purpose, enlightened by reason, had in it nothing of obstinacy; yet resisted those appeals to affection, to interest, or to prejudice, under which so many succumbed.

Within a month after his leaving the Brooklyn, on the 6th of November, 1860, the presidential election was held, and resulted, as had been expected, in the choice of Mr. Lincoln. On the 20th of December South Carolina seceded, and her course was followed within the next six weeks by the other cotton States. In February, 1861, delegates from these States met in convention at Montgomery, Alabama, adopted a constitution, and elected Jefferson Davis to be president of their confederation. On the 18th he was inaugurated, and the new government was thus formally constituted.

Here for a moment the secession movement paused, and Farragut earnestly trusted would stop. Born in a Southern State, and passing his childhood in the extreme Southwest, his relations with both had been severed at too early an age to establish any lasting hold upon his affections; but, though he was to the end carried upon the Navy Register as a citizen of Tennessee, the tenderest and most enduring ties of his life had been formed in Virginia. Nowhere were local bonds stronger, nowhere State pride greater or more justified, than in the famous Commonwealth, which had stood in the center of the line in the struggle for independence, and had given to the nation so many illustrious men from Washington downward. It was impossible that Farragut—who at so early an age, and when attached to no other spot, had married in Norfolk, and thenceforward gone in and out among its people—should be insensible to these influences, or look without grief to a contingency which should force him to sunder all these associations and go forth, on the verge of old age, to seek elsewhere a new home. Nor is it possible to many, however conscious of right, to bear without suffering the alienation and the contempt visited upon those who, in times of keen political excitement, dare to differ from the general passion which sways the mass around them.

Farragut therefore naturally hoped that this bitter trial might be spared him. The Virginian people had taken what seemed then to be a conservative attitude; and, although he was determined to abide by the Union if it were severed by violent action, he was anxious to believe that his home might be saved to him. The Legislature of the State met early in January and recommended all the States to appoint deputies to a peace convention, which accordingly met on the 4th of February; but the propositions made by it were not such as the National Congress could accept. On the 13th of the same month there was assembled at Richmond a State convention, the majority of the delegates to which were Union men, in the then sense of the word in that State. This fact, and the character of some of the speeches made, tended to encourage the belief to which Farragut's wishes led him; but this hope was soon damped by the passage of resolutions affirming the right of secession, and defining the grounds upon which Virginia would be justified in exercising the right. Among these grounds were the adoption of any warlike measures by the United States Government, the recapture of the forts which had been seized by the States already seceded, or any attempt to exact duties from them. True, this was followed during the first week in April by the rejection of a proposition to secede by a vote of eighty-nine to forty-five; but, as Farragut held that the President would be justified in calling out troops when the forts and property of the nation had been violently taken from it, the contrary avowal of the Legislature of his State showed that he might soon be forced to choose between it and the National Government. In that case his mind was fully made up; the choice was painful, but not doubtful. "God forbid," he said, "that I should have to raise my hand against the South!" but the words themselves showed that, however bitter the decision, he was ready to make it. If separation between the sections came peacefully, by mutual consent, he would abide in the only home his manhood had known, and cast his lot thenceforth with the people to whom he was allied and among whom his interests lay; but if the rupture took the form of violent rebellion against the Central Government, whose claims he admitted and to which he owned allegiance, he was prepared to turn his arms even against those who in the other alternative would have been his countrymen. The attitude thus held during those long months of suspense and anxiety was honorable alike to his heart, which responded warmly to the calls of natural affection, and to his conscience, which subordinated the dictates of the heart to his convictions of right; while the unhesitating character of his resolution, amid the uncertainties that unsettled so many men, must be attributed to that habit of preparing for emergencies which characterized his career.

On the 12th of April, 1861, the long period of waiting and watching was brought to an end by the attack upon Fort Sumter. On the 15th President Lincoln issued his proclamation formally announcing the condition of affairs which existed in the seceded States, the defiance of the Central Government, and the seizure of its property. In consequence he called for seventy-five thousand men from the militia of the various States, and avowed clearly that "the first service assigned to the forces hereby called forth will probably be to repossess the forts, places, and property which have been seized from the Union." This was clearly an appeal to arms, provoked finally by the assault upon Fort Sumter, but which the convention then sitting in Richmond had pronounced to be a lawful cause for secession. In the excitement of the hour the Union men, whose attitude toward the more violent party had been almost apologetic, were swept away by the current of feeling, and an ordinance of secession was passed by the convention on the 17th of April, 1861.

During the previous winter Farragut had been residing in Norfolk, unemployed by the Government, but in daily association both with citizens and naval officers; many of whom, like himself, were married and settled there. He and his friends met daily at one of those common rendezvous which are to be found in every small town, and there discussed the news which each day brought of change and excitement. In this way Farragut became acquainted with the views of most of the resident officers, and realized, without being himself swayed by, the influences to which all of them, and especially those of Southern birth, were subjected. With the conservatism common in seamen who have been for long periods separated by their profession from their native places, the great majority of these officers, already men of middle age, could not but feel keen sorrow at the prospect of changes, which would remove them from the navy and separate them from the flag which had hitherto stood to them for country. But, moved by feeling and prejudice, wrought upon by the strong appeals of those they loved, and unfortified by the well-reasoned convictions which made the strength of Farragut, it was equally impossible for the greater part of them to imitate his example. The sense of duty and official honor which they owed to their long training in a generous service stood by them, and few were the cases of men false to trusts actually in their charge; but theirs was not that sense of personal allegiance to the Government which gave the light of the single eye, and enabled Farragut's final decision to be as prompt as it was absolute.

On the 18th of April, the day after the ordinance of secession had been passed, Farragut went as usual to the place of meeting, and saw, immediately upon entering, by the faces of those there, that a great change had passed over the relations between them. He spoke with his usual openness, and expressed his deliberate convictions. He did not believe that the action of the convention represented the sober judgment of the people. The State had been, as he phrased it, "dragooned" out of the Union; and President Lincoln was perfectly justified in calling for troops after the seizure of the forts and arsenals. One of those present remarked impatiently that a person with such sentiments could not live in Norfolk, and this feeling was evidently shared by the bystanders; there was, indeed, some danger, in those excited moments, of personal violence to those who dared gainsay the popular passion. "Very well," replied Farragut, "I can live somewhere else." No time was needed to take a decision already contingently formed, and for executing which he had, with his customary foresight, been accumulating the necessary funds. He at once went to his house and told his wife the time had come for her to decide whether she would remain with her own kinsfolk or follow him North. Her choice was as instant as his own, and that evening they, with their only son, left Norfolk, never to return to it as their home. Mrs. Farragut's sister and her young family accompanied them in the steamer to Baltimore. Upon reaching the latter city they found it also boiling over with excitement. The attack upon the Massachusetts troops had just taken place, and the railroad bridges over the Susquehanna were then burning. The usual means of communication being thus broken off, Farragut and his party had to take passage for Philadelphia in a canal boat, on which were crowded some three hundred passengers, many of them refugees like themselves. It is a curious illustration of the hardships attending a flight under such exigency, even in so rich a country as our own, that a baby in the company had to be fed on biscuit steeped in brandy for want of proper nourishment.

From Philadelphia the journey to New York was easy, and Farragut there settled his family in a small cottage in the village of Hastings, on the Hudson River. Here he awaited events, hoping for employment; but it is one of the cruel circumstances attending civil strife that confidence is shaken, and the suspicions that arise, however unjust, defy reason and constrain the Government to defer to them. No man could have given stronger proof than Farragut had of his perfect loyalty; but all shades of opinion were known to exist among officers of Southern origin, even when they remained in the service, and there were those who, though refusing to follow the South, would willingly have avoided striking a blow against the seceding States. Men were heard to say that they would not go with their State, but neither would they fight against her; or that they would remain in the navy, but seek employment that might spare them the pain of taking part in such a contest. These illogical positions were soon abandoned as the spirit of war gained more and more hold upon the feelings of men, but for Farragut they never existed after the first blow was struck. Through whatever struggles with himself he may have passed in the earlier stages of the secession movement, his decision, when reached, admitted no half-measures, nor halted between two opinions. "He stood on no neutral ground, he longed to take an active part in the war." Nevertheless, the Government could not at once accept, as a title to full and implicit confidence, even the sacrifice of home and life-long associations which he had made to the cause of the Union. If given any duty, a man of Farragut's rank and attainments must needs have one involving much responsibility, failure in which would involve not only himself but those who had employed him. The cry of treachery was sure to follow, and prudent officers of Southern birth found it advisable to decline employments where they foresaw that delays were unavoidable, because they felt that what might be explained in the case of a Northern man would in them be stamped by public opinion as the result of disaffection. In Hastings and its neighborhood the most grotesque suspicions were spread concerning the Southern captain who had thus come to dwell among them, and who, for conscience and country, had given up more than had been demanded of those who thus distrusted him. Time was needed to allow men's minds to reach a more reasonable frame, and for the Government itself to sift and test, not merely the fidelity, but the heartiness and the probable capacity of the officers at its command.

Farragut's first employment was as a member of a board to recommend officers for retirement from active service, under an act approved August 3, 1861. The object of this act was to assist the Department in the discrimination necessary to be made between the competent and those disabled by years or infirmity, for up to that time there had been no regular system of retirement, and men were retained on the active list past the period of efficiency, because no provision for removing them existed. The duty, though most important with war actually existing, was delicate and trying, and far from consonant to Farragut's active, enterprising character. More suitable employment was, however, fast approaching

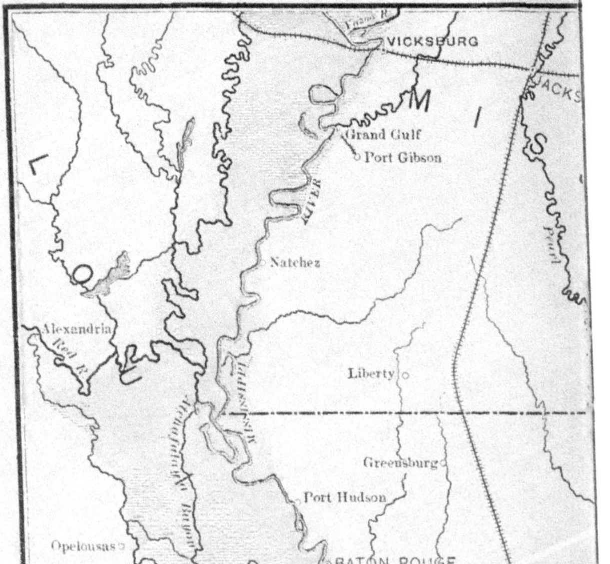

Bradley & Poates Engrs, N.Y.

Scene of Farragu

OPERATIONS, 1862–1864.

CHAPTER VII.

THE NEW ORLEANS EXPEDITION.

1862.

THE necessity of controlling the Mississippi valley had been early realized by the United States Government. In its hands the great stream would become an impassable barrier between two large sections of the Southern Confederacy; whereas in the possession of the latter it remained a link binding together all the regions through which it flowed, or which were penetrated by any of its numerous tributaries. The extensive territory west of the river also produced a large part of the provisions upon which depended the Southern armies, whose main field of action was, nevertheless, on the eastern side. In a country habitually so unprepared for war as is the United States, and where, of course, such a contingency as an intestine struggle between the sections could not have been provided for, there seemed room to hope that the national forces might by rapid action seize the whole course of the river, before the seceding States were able to take adequate measures for its defense. The Government had the support of that part of the country which had received the largest manufacturing development, and could, therefore, most quickly prepare the material for war, in which both sides were lamentably deficient; and, what was yet more important, it possessed in the new navy built since 1855 an efficient weapon to which the South had nothing to oppose. The hope was extravagant and doomed to disappointment; for to overrun and hold so extensive a territory as the immediate basin of the Mississippi required a development of force on the one side and a degree of exhaustion on the other which could not be reached so early in the war. The relative strengths, though unequal, were not yet sufficiently disproportioned to enable the gigantic work to be accomplished; and the principal result of an effort undertaken without due consideration was to paralyze a large fraction of a navy too small in numbers to afford the detachment which was paraded gallantly, but uselessly, above New Orleans. Nor was this the worst; the time thus consumed in marching up the hill in order at once to march down again threw away the opportunity for reducing Mobile before its defenses were strengthened. Had the navy been large enough, both tasks might have been attempted; but it will appear in the sequel that its scanty numbers were the reason which postponed the attack on Mobile from month to month, until it became the most formidable danger Farragut ever had to encounter.

Despite the extensive sea-coast of the United States and the large maritime commerce possessed by it at the opening of the war, the navy had never, except for short and passing intervals, been regarded with the interest its importance deserved. To this had doubtless contributed the fixed policy of the Government to concentrate its attention upon the internal development of the country, and to concern itself little with external interests, except so far as they promoted the views of that section which desired to give extension to slaveholding territory. The avoidance of entangling alliances had become perverted to indifference to the means by which alone, in the last resort, the nation can assert and secure control in regions outside its borders, but vitally affecting its prosperity and safety. The power of navies was therefore, then as now, but little understood. Consequently, when the importance of the Mississippi Valley was realized, as it immediately was, there was but one idea as to the means of controlling it, and that was by a land invasion from the great Western and Northwestern States. To this a navy was indeed to be adjoined, but in a manner so distinctly subsidiary that it was, contrary to all custom, placed under the orders of the commander-in-chief of the Western army, and became simply a division of the land forces. From this subordinate position it was soon raised by its own intrinsic value and the logic of facts; but the transient experience is noteworthy, because illustrating the general ignorance of the country as to the powers of the priceless weapon which lay ready, though unnoticed, to its hand.

Happily, in the Navy Department itself juster views prevailed; and the general indifference permitted it at least one compensation—to follow its own ways. The Secretary himself was not a professional man, though he had had official connection with the service in the past; but most fortunately there was called to his assistance one who had been for eighteen years in the navy, had passed while in it to the command of mail steamers, and only five years before the war had resigned and entered civil life. This gentleman, Mr. Gustavus V. Fox, thus combined with business experience and an extensive acquaintance with naval officers the capacities of a seaman. He knew what ships could do and what they could not; but to this common knowledge of sea officers, gained by the daily habit of sea life, he had added the results of study and reflection upon events passing elsewhere than under his own observation. The experiences of the allied navies in the Crimean War had convinced him that, if the wooden sides of ships could not be pitted in prolonged stand-up fight against the stone walls of fortresses, they were capable of enduring such battering as they might receive in running by them through an unobstructed channel. This conviction received support by the results of the attacks upon Hatteras Inlet and Port Royal. He might, indeed, have gone much further back and confirmed his own judgment as a seaman by the express opinion of an eminent soldier. Nearly a hundred years before, Washington, at the siege of Yorktown, had urged the French Admiral De Grasse to send vessels past Cornwallis's works to control the upper York River, saying: "I am so well satisfied by experience of the little effect of land batteries on vessels passing them with a leading breeze that, unless the two channels near Yorktown should be found impracticable by obstructions, I should have the greatest confidence in the success of this important service."*

In this conviction of Mr. Fox's lay the inception

* *Washington's Letters*, October 1, 1781.

of the expedition against New Orleans. It was, in his view, to be a purely naval attack. Once over the bar at the mouth of the river, the channel as far as the city had no natural obstruction, was clearly defined, and easily followed, by day or night, without a pilot. The heavy current of the early spring months, while it would retard the passage of the ships and so keep them longer under fire, would make it difficult for the enemy to maintain in position any artificial barrier placed by him. The works to be passed—the seaward defenses of New Orleans, Forts Jackson and St. Philip—were powerful fortifications; but they were ultimately dependent upon the city, ninety miles above them, for a support which could come only by the river. A fleet anchored above the forts lay across their only line of communication, and when thus isolated, their fall became only a question of time. The work proposed to the United States Navy was, therefore, to turn the forts by passing their fire, seize their line of communications—the upper river—and their base, New Orleans, and then to give over the latter to the army, which engaged to furnish a force sufficient to hold the conquest.

Having first taken the necessary, but strictly preliminary, step of seizing as a depot Ship Island, in Mississippi Sound, about a hundred miles from the mouth of the river, Mr. Fox's proposition, which had been adopted by the Secretary of the Navy, was submitted to the President. Mr. Lincoln, himself a Western man, unfamiliar with maritime matters and engrossed with the idea of invasion from the north, was disposed to be incredulous of success; but with his usual open-mindedness consented to a full discussion before him by experts from both services. A meeting was therefore held with General McClellan at his headquarters. There were present, besides the President, the Secretary of the Navy, Mr. Fox, and Commander David D. Porter, who had recently returned from service off the mouth of the Mississippi. The antecedents of General McClellan were those of an officer of the engineers, who are generally disposed to exaggerate the powers of forts as compared with ships, and to contemplate their reduction only by regular approaches; just as an officer of the line of the army, looking to the capture of a place like New Orleans, will usually and most properly seek first a base of operations, from which he will project a campaign whose issue shall be the fall of the city. To this cause was probably due the preference observed by the Navy Department to exist in army circles, for an attack upon Mobile first. Being close to the sea, which was completely under the control of the navy, the necessary land operations would begin under far more favorable conditions, and could be more easily maintained than in the alluvial soil of the Mississippi delta. McClellan, who was an accomplished master of his profession in all its branches, received at first the impression that regular military operations against New Orleans by way of the river were being proposed to him, and demurred; but, on learning that the only demand was for a force to hold the city and surroundings in case of success, he readily consented to detail ten or fifteen thousand troops for the purpose. Though more hazardous, the proposition of the Navy Department was in principle strategically sound. The key of the position was to be struck for at once, and the outlying defenses were expected then to fall by the severance of their communications. The general might have his own opinion as to the power of the navy to carry out the proposed passage of the forts, and as to whether its coal, when once above, would outlast the endurance of the hostile garrisons; but those were points upon which the Navy Department, which undertook the risk, might be presumed to have more accurate judgment than himself.

The conference, which was held about the middle of November, 1861, resulted in the adoption of Mr. Fox's plan in its main outlines; but with an important addition, which threatened at one time to become a very serious modification. Commander Porter suggested that the naval vessels should be accompanied by a mortar flotilla, to subdue the fire of the forts by bombardment, and so to allow the fleet to pass without risk, or with risk much diminished. This proposition approved itself to the engineer instincts of McClellan, and was adopted. The general then designated Major Barnard, of the Engineer Corps, to represent him in adjusting the details of the expedition. Barnard also took strong ground in favor of the mortars, and to this added the opinion—in which Porter concurred—that the forts should be not merely bombarded, but reduced before the passage. He summed up his conclusions in the following perfectly clear words: "To pass those works (merely) with a fleet and appear before New Orleans is merely a raid—no capture. New Orleans and the river can not be held until communications are perfectly established." The assertion of the last sentence can not be denied; it admits of no difference of opinion. The point in dispute between the two arguments was not this, but whether the fall of the city, which had no local defenses, would entail that of the forts, and so open the communications. Mr. Fox strongly held that it would; but although he stuck to his opinion, he had a deservedly high estimate of Porter's professional ability—so much so that, had the latter's rank justified, he would have urged him for the command of the expedition. In this doubtful state of the argument, it will be seen of how great importance was the choice of the officer to be put in charge of the whole undertaking. Had he also taken the view of Barnard and Porter in favor of the more cautious, but—as it proved—more dangerous course, it could scarcely have failed that Fox would have been overruled.

The nomination of this officer could not be longer deferred. Secrecy and rapidity of action were large elements in the hoped-for achievement, and secrecy depends much upon the length of time the secret must be kept. Among the officers whose length of service and professional reputation indicated them as suitable for the position, there was little to guide the department to the man who would on emergency show the audacity and self-reliance demanded by the intended operations. The action proposed, though it falls within the limits of the methods which history has justified, and has, therefore, a legitimate place in the so-called science of war, was, nevertheless, as the opinions of Barnard and Porter show, contrary to the more usual and accepted practice. It disregarded the safeguards commonly insisted upon, overleaped the successive steps by which military achievement ordinarily advances to its end, and, looking only to the exceptional conditions, resorted fearlessly to exceptional methods. For such a duty the department needed a man of more than average determination and vigor.

Farragut's name was necessarily among those considered; but the final choice appears to have been determined by the impression made upon Mr. Fox, and through him upon the department, by his course in leaving Norfolk at the time and in the way he did. This, Fox argued, showed "great superiority of character, clear perception of duty, and firm resolution in the performance of it." His conspicuous ability was not then recognized, could not be until revealed by war; but it was evident that he stood well above the common run of simply accomplished officers. Still, further tests were required; in a matter of so much importance the department had need to move warily. That Farragut was faithful could not be doubted; but was his heart so far in the contest that he could be depended upon to exert his abilities to the full? Commander Porter was ordered to go to New York on duty connected with the mortar flotilla, and while there to make an opportunity to visit Farragut. There had been, as is known, a close relation between the two families, and to him Farragut was likely to show how hearty he was in the cause. Porter's account was most favorable, and it then remained only to judge whether he was in sympathy with the military plan of the proposed expedition.

For this object Farragut was ordered to report at the department, and Fox undertook to meet him at the train and talk over the matter informally. He arrived in Washington on the 21st of December, was met as arranged, and taken to the house of the Postmaster-General, Montgomery Blair. The latter was brother-in-law to Fox, and the three breakfasted together. "After breakfast, Fox laid before Farragut the plan of attack, the force to be employed, and the object to be attained, and asked his opinion. Farragut answered unhesitatingly that it would succeed. Fox then handed him the list of vessels being fitted out, and asked if they were enough. Farragut replied he would engage to run by the forts and capture New Orleans with two thirds the number. Fox told him more vessels would be added, and that he would command the expedition. Farragut's delight and enthusiasm were so great that when he left us Fox asked if I did not think he was too enthusiastic. I replied I was most favorably impressed with him, and sure he would succeed." * There could be no question, at any rate, that his whole heart was in the war and in the expedition; whether he would rise equal to his task still remained to be seen. He said, however, frankly, that had he been previously consulted, he would have advised against the employment of the mortar flotilla. He had no faith in the efficacy of that mode of attack since his observations of the results at San Juan de Ulloa, twenty-three years before. He was convinced that the fleet could run by the forts, and anticipated nothing but delay from the bombardment. Nevertheless, since the arrangements had been made, he was willing to give the bombs a trial. "He was never profuse in promises," writes Mr. Welles, the Secretary of the Navy, "but he felt complimented that he was selected, and I saw that in modest self-reliance he considered him-

* Montgomery Blair, in *The United Service*, January, 1881.

self equal to the emergency and to the expectation of the Government."* To his home he wrote: "Keep your lips closed and burn my letters, for perfect silence is to be observed—the first injunction of the Secretary. I am to have a flag in the Gulf, and the rest depends upon myself. Keep calm and silent. I shall sail in three weeks."

On the 23d of December, 1861, Farragut received preparatory orders, and on the 9th of the following January was formally appointed to command the Western Gulf Blockading Squadron; the limits of which, on the coast of the Confederacy, were defined as from St. Andrew's Bay to the mouth of the Rio Grande. The coasts of Mexico and Yucatan were also embraced in them. The steam sloop-of-war Hartford was selected for his flag-ship. On the 20th of January final orders were issued to him. These were somewhat discreetly worded, and, literally understood, must be conceded to take from the department the credit of boldly adhering to, and assuming the responsibility of, the original plan—a credit Mr. Welles seems desirous to claim. "When you are completely ready," they read, "you will collect such vessels as can be spared from the blockade, and proceed up the Mississippi River *and reduce the defenses* which guard the approaches to New Orleans, *when* you will appear off that city and take possession of it under the guns of your squadron." Understood according to the plain meaning of the words, these orders prescribed the reduction of the works as a condition precedent to appearing off the city, and so recur to the fears expressed by both Barnard and Porter as to the consequences of

* Gideon Welles, in the *Galaxy*, November, 1871.

leaving the forts unreduced. There is not in them even "the latitude and discretion in the employment of the means placed under his command" which Mr. Welles claimed.* Had Farragut, after leaving the forts unreduced, as he did, met with serious disaster, it can scarcely be doubted that the phrase quoted would have been used to acquit the Government.

The steam-sloop Hartford, upon which Farragut now hoisted his flag, and in which he continued throughout the war, was a nearly new vessel, having sailed on her first cruise to China in the summer of 1859. She belonged to the early period of the transition from sails to steam for the motive power of vessels; the steam being regarded as auxiliary only, and giving her a speed of but eight knots per hour, while the spars and sail area were those of a full-rigged ship. The deficiency of horse-power was a serious drawback in such an operation as passing forts, especially when, as in the Mississippi, the current was strong and always adverse to vessels ascending the river. The Hartford had, on the other hand, a powerful battery of the best existent type. She carried twenty-two Dahlgren nine-inch shell guns, eleven on each side; and, owing to the lowness of the river banks, these guns would be on a level with or even above those in the lower tier of the batteries opposed to her. The Pensacola, Brooklyn, and Richmond were vessels of the same type as the Hartford, and built at the same time.

On the 2d of February, 1862, the Hartford sailed from Hampton Roads, and on the 20th reached Ship Island. The following day Farragut took over the

* Gideon Welles, in the *Galaxy*, December, 1871.

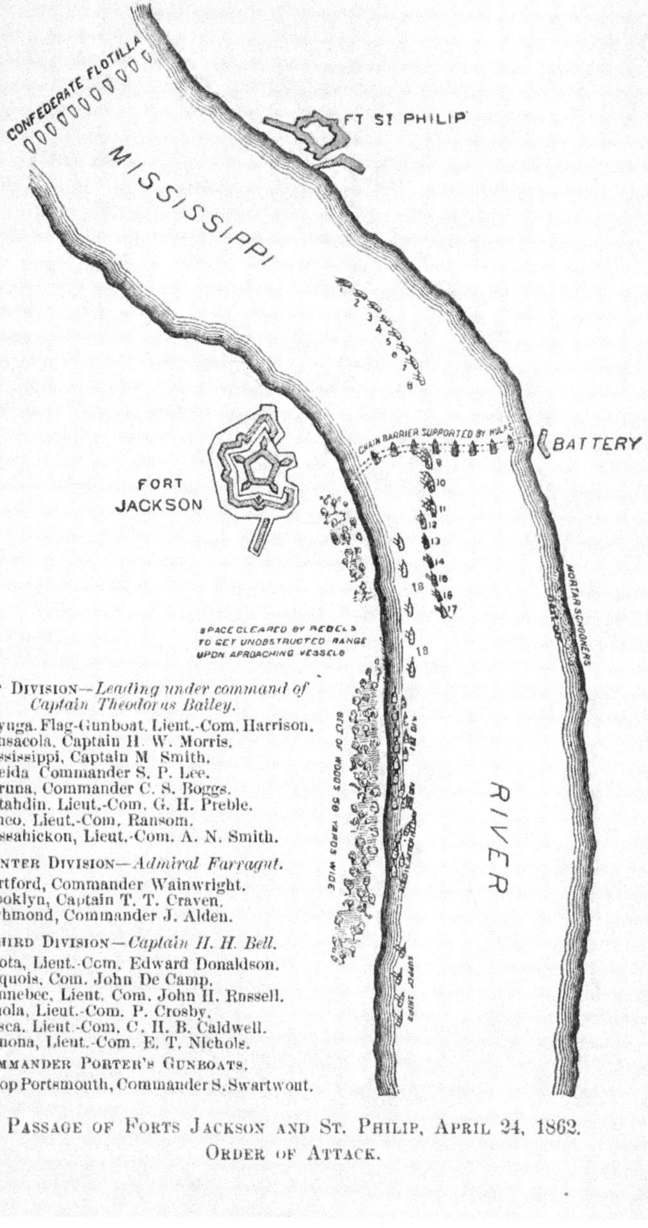

FIRST DIVISION—*Leading under command of Captain Theodorus Bailey.*

1. Cayuga. Flag-Gunboat. Lieut.-Com. Harrison.
2. Pensacola, Captain H. W. Morris.
3. Mississippi, Captain M. Smith.
4. Oneida Commander S. P. Lee.
5. Varuna, Commander C. S. Boggs.
6. Katahdin. Lieut.-Com. G. H. Preble.
7. Kineo. Lieut.-Com. Ransom.
8. Wissahickon, Lieut.-Com. A. N. Smith.

CENTER DIVISION—*Admiral Farragut.*

9. Hartford, Commander Wainwright.
10. Brooklyn, Captain T. T. Craven.
11. Richmond, Commander J. Alden.

THIRD DIVISION—*Captain H. H. Bell.*

12. Sciota, Lieut.-Com. Edward Donaldson.
13. Iroquois, Com. John De Camp.
14. Kennebec, Lieut. Com. John H. Russell.
1 . Pinola, Lieut.-Com. P. Crosby.
16. Itasca. Lieut.-Com. C. H. B. Caldwell.
17. Winona, Lieut.-Com. E. T. Nichols.

18 COMMANDER PORTER'S GUNBOATS.

19. Sloop Portsmouth, Commander S. Swartwout.

PASSAGE OF FORTS JACKSON AND ST. PHILIP, APRIL 24, 1862.
ORDER OF ATTACK.

command of his district and squadron from Flag Officer McKean, who up to that time had had charge of both the East and West Gulf. None of the other vessels of the expedition were yet there; but they came in one by one and were rapidly assembled at the Southwest Pass, then the principal entrance to the river. Much difficulty was encountered in getting the heavier ships over the bar, two weeks' work being needed to drag the Pensacola inside; but on the 7th of April she floated in the river, and Farragut found his force complete. It then consisted, independently of the steamers attached to the mortar flotilla, of four steam sloops-of-war of about two thousand tons each, three of half that size, one large side-wheel ship-of-war, the Mississippi, of seventeen hundred tons, and nine gun-boats of five hundred. The latter had been hurriedly built to meet the special exigencies of this war, and were then commonly known as the "ninety-day" gunboats. Each carried one eleven-inch shell-gun and one thirty-pounder rifle. The aggregate batteries of the seventeen vessels composing the squadron, excluding some light brass pieces, amounted to one hundred and fifty-four cannon, of which one hundred and thirty-five were thirty-two pounders or above.

The two forts which constituted the principal defenses of New Orleans against a naval attack from the sea were at Plaquemine Bend, about twenty miles above the Head of the Passes; by which name is known the point where the main stream of the Mississippi divides into several channels, called passes, through which its waters find their way to the Gulf. The river, whose general course below New Orleans is southeast, turns at Plaquemine Bend northeast for a mile and three-quarters, and then resumes its previous direction. The heavier of the two works, Fort Jackson, is on the right bank, at the lower angle of the Bend. It was a casemated brick structure, pentagonal in form, carrying in barbette over the casemates twenty-seven cannon of and above the size of thirty-two pounders, besides eleven twenty-four pounders. In the casemates were fourteen of the latter caliber. Attached to this fort, but below it, was a water battery carrying half a dozen heavy cannon. Fort St. Philip was nearly opposite Fort Jackson, but somewhat below it, so as to command not only the stream in its front, but also the stretch down the river, being thus enabled to rake vessels approaching from below before they came abreast. It comprised the fort proper and two water batteries, which together mounted forty-two guns. The sites of these fortifications had been skillfully chosen; but their armaments, though formidable and greatly superior to those of the fleet—regard being had to the commonly accepted maxim that a gun ashore is equivalent to four afloat—were not equal to the demands of the situation or to the importance of New Orleans. Out of a total of one hundred and nine pieces,* of which probably over ninety could be used against a passing fleet, fifty-six, or more than half, were of the very old and obsolete caliber of twenty-four pounders.

This inadequate preparation, a year after the attack upon Fort Sumter and the outbreak of hostilities, is doubtless to be attributed to surprise. The Southern authorities, like those of the National

* There were some guns bearing inland and some flanking howitzers, besides those already enumerated.

Government, were firmly possessed with the idea that the Mississippi, if subdued at all, must be so by an attack from the north. Despite the frequency of spies and treason along the border line of the two sections, the steps of the Navy Department were taken so quietly, and followed so closely upon the resolve to act, that the alarm was not quickly taken; and when intimations of attack from the sea did filter through, they had to encounter and dislodge strong contrary preoccupations in the minds of the Southern leaders. Only the Confederate general commanding the military division and his principal subordinates seem to have been alive to the danger of New Orleans, and their remonstrances had no effect. Not only were additional guns denied them and sent North, but drafts were made on their narrow resources to supply points considered to be in greater danger. A striking indication of the prepossessions which controlled the authorities at Richmond was elicited by Commodore Hollins, of the Confederate Navy. That gallant veteran was ordered to take to Memphis several of the rams extemporized at New Orleans. He entreated the Navy Department to allow him to remain, but the reply was that the main attack upon New Orleans would be from above, not from below. After the fleet entered the river he telegraphed from Memphis for permission to return, but received the answer that the proposition was wholly inadmissible. Before the Court of Inquiry upon the loss of New Orleans, he testified that the withdrawal of his ships was the chief cause of the disaster.*

* *Official Records of the War of the Rebellion*, Series I, vol. vi, p. 610.

While the heavy ships were being dragged over the bar at the Southwest Pass, the mortar flotilla had entered the river under the command of Commander Porter. No time was avoidably lost, though there were inevitable delays due to the magnitude of the preparations that in every quarter taxed the energies of the Government. On the 16th of April, less than ten days after the Pensacola got safely inside, the fleet was anchored just out of range of the forts. On the 18th the mortar vessels were in position, and at 10 A. M. the bombardment by them began, continuing throughout the succeeding days till the passage of the fleet, and being chiefly directed upon Fort Jackson. From daylight to dark a shell a minute was fired, and as the practice was remarkably good a great proportion of these fell within the fort. As Farragut had predicted, they did not in the course of six days' bombardment do harm enough to compel a surrender or disable the work; but they undoubtedly harassed the garrison to an extent that exercised an appreciable effect upon the fire of Jackson during the passage.

While the bombardment was progressing, the lighter vessels of the squadron were continuously engaged by detachments in protecting the mortar flotilla, steaming up above it and drawing upon themselves the fire of the forts. A more important duty was the removal of the obstructions that the enemy had thrown across the river, below the works, but under their fire. Opinions differed, both in the United States squadron and in the counsels of the enemy, as to the power of the ships to pass the forts; but it was realized on both sides that any barrier to their passage which should force them to stop under fire, or should throw confusion into their order, would materially increase the chances against them. Whatever the blindness or neglect of the Confederate Government, the Confederate officers of the department had not been remiss in this matter. The construction of a floating barrier had early engaged their attention, and, despite the difficulties presented by so rapid a current, a formidable raft had been placed early in the winter. It consisted of cypress logs forty feet long and four or five feet in diameter, lying lengthwise in the river, with an interval of three feet between them to allow drift to pass. The logs were connected by two and a half inch iron cables, stretching underneath from one side of the stream to the other; and the whole fabric was held up against the current by some thirty heavy anchors and cables. So long as it stood, this constituted a very grave difficulty for an attacking fleet; but the water was deep and the holding ground poor, so that even under average conditions there was reason to fear its giving way. The fleet arrived in the early spring, the season when the current, swollen by the melting snows about the head waters of the Mississippi and its tributaries, is at its strongest; and in 1862 the spring rise was greater than for many years. In February the raft began to show signs of yielding under the pressure of the drift wood accumulating on it from above, and on the 10th of March the cables had parted, the sections on either side being swept against the banks and leaving about a third of the river open. The gap was filled by anchoring in it eight heavy schooners of about two hundred tons burden. They were joined together as the cypress logs had been, but with lighter chains, probably because no heavy ones were at hand; and, as a further embarrassment to the assailants, their masts were unstepped and allowed to drag astern with the rigging attached, in the hopes that by fouling the screws the ascending vessels might be crippled.

This central barrier of schooners was not intrinsically strong, but it was not to be despised, considering the very moderate speed possessed by the ships and the strength of the current which they had to stem. It was doubtful whether they could break through with so little loss of way as to produce no detention; and the mere presence of so many hulls on a dark night and under the added gloom of the battle's smoke was liable to increase a confusion which could redound only to the advantage of the defense. It became necessary, therefore, to remove the schooners in whole or in part. This was effected in a very daring manner by two gunboats, the Itasca and Pinola, Captains Caldwell and Crosby; the fleet captain, Henry H. Bell, an officer in whom Farragut had the most unbounded confidence, being placed in command of both. The work had to be done, of course, within range of the hostile batteries, which, through some culpable negligence, failed to molest it. The Pinola carried an electrician with a petard, by which it was hoped to shatter the chains. This attempt, however, failed, owing to the wires of the electrical battery parting before the charge could be exploded. The Itasca, on the other hand, ran alongside one of the schooners and slipped the chains; but, unfortunately, as the hulk was set adrift without Captain Caldwell being notified, and the engines of the gunboat were going ahead with the helm a-port, the two vessels turned inshore and ran aground under fire of the forts. In this critical position the Itasca remained for some time, until the Pinola could be recalled to her assistance; and then several attempts had to be made before she finally floated. Caldwell then did an exceedingly gallant thing, the importance of which alone justified, but amply justified, its temerity. Instead of returning at once to the squadron, satisfied with the measure of success already attained, he deliberately headed up the river; and then, having gained sufficient ground in that direction to insure a full development of his vessel's speed, he turned and charged full upon the line of hulks. As she met the chains, the little vessel rose bodily three or four feet from the water, sliding up on them and dragging the hulks down with her. The chains stood the strain for an instant, then snapped, and the Itasca, having wrought a practicable breach, sped down to the fleet.

While these various accessory operations were going on, Admiral Farragut's mind was occupied with the important question of carrying out the object of his mission. The expedient of reducing or silencing the fire of the enemy's forts, in which he himself had never felt confidence, was in process of being tried; and the time thus employed was being utilized by clearing the river highway and preparing the ships to cut their way through without delay, in case that course should be adopted. Much had been done while at the Head of the Passes, waiting for the Pensacola to cross the bar; but the work was carried on unremittingly to the last moment. The loftier and lighter spars of all the vessels had already been sent ashore, together with all unnecessary encumbrances, several of the gunboats having even unstepped their lower masts; and the various ordinary precautions, known to seamen under the name of "clearing ship for action," had been taken with reference to fighting on anchoring ground. These were particularized in a general order issued by the admiral, and to them he added special instructions, rendered necessary by the force of the current and its constancy in the same direction. "Mount one or two guns on the poop and top-gallant forecastle," he said; "in other words, be prepared to use as many guns as possible ahead and astern to protect yourself against the enemy's gunboats and batteries, bearing in mind that you will always have to ride head to the current, and can only avail yourself of the sheer of the helm to point a broadside gun more than three points (thirty-four degrees) forward of the beam. . . . Trim your vessel also a few inches by the head, so that if she touches the bottom she will not swing head down the river," which, if the stern caught the bottom, would infallibly happen, entailing the difficult manœuvre and the perilous delay of turning round under the enemy's fire in a narrow river and in the dark. The vessels generally had secured their spare iron cables up and down their sides in the line of the boilers and engines; and these vital parts were further protected by piling around them hammocks, bags of sand or ashes, and other obstructions to shot. The outsides of the hulls were daubed over with Mississippi mud, to be less easily discerned in the dark; while the decks were whitewashed, so as to throw in stronger relief articles lying upon them which needed to be quickly seen.

Having given his general instructions, the flag officer could intrust the details of preparation to his subordinates; but no one could relieve him of the momentous decision upon which the issues of the campaign must turn. The responsibility of rejecting one course of action and adopting another was his alone; and as has already been remarked, the wording of the department's order, literally understood, imposed upon him the task of reducing the forts before approaching the city. The questions involved were essentially the same as those presented to every general officer when the course of a campaign has brought him face to face with a strong position of the enemy. Shall it be carried by direct attack, and, until so subdued, arrest the progress of the army? or can it be rendered impotent or untenable by severing its communications and by operations directed against the district in its rear, which it protects, and upon which it also depends? The direct attack may be by assault, by investment, or by regular siege approaches; but whatever the method, the result is the same—the assailant is detained for a longer or shorter time before the position. During such detention the post fulfills its mission of securing the region it covers, and permits there the uninterrupted prosecution of the military efforts of every character which are designed to impede the progress of the invader.

To such cases no general rule applies; each turns upon particular conditions, and, although close similarities may exist between various instances, probably no two are entirely identical. It is evident, however, that very much will depend upon the offensive power shut up in the position under consideration. If it be great walled town, such as are found on the Continent of Europe, behind whose defenses are sheltered numerous troops, the assailant who advances beyond it thereby exposes his communications to attack; and, to guard against this danger, must protect them by a force adequate to hold the garrison in check. If, again, there be but a single line by which the communications can be maintained, by which supplies and re-enforcements can go forward, and that line passes close under the work and is commanded by it, the garrison may be small, incapable of external action, and yet may vitally affect the future operations of the venturesome enemy who dares to leave it unsubdued behind him. Such, to some extent, was the Fort of Bard, in the narrow pass of the Dora Baltea, to Napoleon's crossing of the St. Bernard in 1800; and such, to some extent, would be Forts Jackson and St. Philip to Farragut's fleet after it had fought its way above. The Mississippi was the great line of communication for the fleet; no other was comparable to it—except as a by-path in a mountain is comparable to a royal highway—and the forts commanded the Mississippi. Their own offensive power was limited to the range of their guns; their garrisons were not fitted, either by their number or their aptitudes, for offensive action upon the water; but so long as their food and ammunition lasted, though an occasional vessel might run by them, no steady stream of supplies, such as every armed organization needs, could pass up the Mississippi. Finally, though the garrison could not move, there lay behind or under the forts a number of armed vessels, whose precise powers were unknown, but concerning which most exaggerated rumors were current.

The question, therefore, looming before Farragut was precisely that which had been debated before the President in Washington; precisely that on which Fox had differed from Porter and Barnard. It was, again, closely analogous to that which divided Sherman and Grant when the latter, a year after Farragut ran by the forts, made his famous decision to cut adrift from his communications by the upper Mississippi, to march past Vicksburg by the west bank of the river, to cross below the works, and so cut off the great stronghold of the Mississippi from the country upon which it depended for food and re-enforcements.* But as Grant's decision rested upon

* The following is Grant's account of a matter which, but for Sherman's own zeal in proclaiming the merits of his commander-in-chief, would probably have always remained unknown. It would be difficult to find a closer parallel to the difference of judgment existing between Farragut and Porter at New Orleans: "When General Sherman first learned of the move I proposed to make, he called to see me about it. I was seated on the piazza, engaged in conversation with my staff, when he came up. After a few moments' conversation, he said he would like to see me alone. We passed into the house together and shut the door after us. Sherman then expressed his alarm at the move I had ordered, saying that I was putting myself voluntarily in a position which an enemy would be glad to manœuvre a year—or a long time—to get me in. I was going into the enemy's country, with a large river behind me, and the enemy holding points strongly fortified above and below. He said that it was an axiom in war that when any great body of troops moved against an enemy they should do so from a base of supplies which they would guard as the apple of the eye, etc. He pointed out all the difficulties that might be encountered in the campaign proposed, and stated in turn what would be the true campaign to make. This was, in substance, to go back until high ground could be reached on the east bank of the river, fortify there and establish a depot of supplies, and move from there, being always prepared to fall back

a balance of arguments applicable to the problem before him, so did Farragut's upon a calculation of the risks and advantages attendant, respectively, upon the policy of waiting for the forts to fall, or of speeding by them to destroy the resources upon which they depended.

The reasons in favor of waiting for the fall of the works were ably presented by Commander Porter in a paper which he asked to have read in a council of commanding officers of the fleet, assembled on board the flag-ship on the third day of the bombardment, April 20. Farragut was already familiar with the arguments on both sides, and Porter's paper can be regarded only as an expression of views already uttered, but now invested with a formality becoming the seriousness of the occasion. In its finality it has somewhat the character of a protest, though indirect and couched in perfectly becoming language, against a decision which Farragut had now reached and which Porter had always combated. The latter

---

upon it in case of disaster. I said this would take us back to Memphis. Sherman then said that was the very place he should go to, and would move by railroad from Memphis to Granada. To this I replied, the country is already disheartened over the lack of success on the part of our armies, . . . and if we went back so far as Memphis, it would discourage the people so much that bases of supplies would be of no use; neither men to hold them nor supplies to put in them would be furnished. The problem was to move forward to a decisive victory, or our cause was lost. . . . Sherman wrote to my adjutant-general embodying his views of the campaign that should be made, and asking him to advise me at least to get the views of my generals upon the subject. Rawlins showed me the letter, but I did not see any reasons for changing my plans."—*Personal Memoirs of U.S. Grant*, vol. i, p. 542 (note).

does not appear to have doubted the ability of the fleet to pass the works, but he questioned the utility and expediency of so doing. His words were as follows: *

"The objections to running by the forts are these: It is not likely that any intelligent enemy would fail to place chains across above the forts, and raise such batteries as would protect them against our ships. Did we run the forts we should leave an enemy in our rear, and the mortar vessels would have to be left behind. We could not return to bring them up without going through a heavy and destructive fire. If the forts are run, part of the mortars should be towed along, which would render the progress of the vessels slow against the strong current at that point. If the forts are first captured, the moral effect would be to close the batteries on the river and open the way to New Orleans; whereas, if we don't succeed in taking them, we shall have to fight our way up the river. Once having possession of the forts, New Orleans would be hermetically sealed, and we could repair damages and go up on our own terms and in our own time. . . . Nothing has been said about a combined attack of army and navy. Such a thing is not only practicable, but, if time permitted, should be adopted. Fort St. Philip can be taken with two thousand men covered by the ships, the ditch can be filled with fascines, and the wall is easily to be scaled with ladders. It can be attacked in front and rear."

In summoning his captains to meet him on this occasion, Farragut had no idea of calling a council-

---

* The paper being long, only those parts are quoted which convey the objections to running by.

of-war in the sense which has brought that name into disrepute. He sent for them, not because he wanted to make up his mind, but because it was made up, and he wished at once to impart to them his purposes and receive the benefit of any suggestion they might make. Bell, the chief-of-staff, who was present, has left a memorandum of what passed, which is interesting as showing that the members were not called to express an opinion as to the propriety of the attack, but to receive instructions as to the method, on which they could suggest improvements.

"April 20, 10 A. M. Signal was made for all captains commanding to repair on board the flag-ship. All being present except the three on guard to-day, viz., Commander De Camp and Lieutenants-Commanding Nichols and Russell, the flag-officer unfolded his plan of operations, assigning the places for every vessel in the fleet in the attack, and exhibited his charts of the river and of the forts. Some discussion was had thereupon, and Commander Alden read a written communication to the flag-officer from Commander Porter at his request, expressing his views as to the operation against the forts. Having read them, Commander Alden folded up the paper and returned it to his pocket, whereupon I suggested the propriety of the document being left with the flag-officer, and the paper was accordingly left in his hands. It was therein stated that the boom being a protection to the mortars against attacks of all kinds from above, the boom should not be destroyed until the forts were reduced. Upon this the flag-officer remarked that the commander had this morning assented to the propriety of the boom being broken to night—which I heard—and, again, that the fleet should not go above the forts, as the mortar fleet would be left unprotected. The flag-officer thought the mortars would be as well protected above as below the forts, and that co-operation with the army, which entered into the plans of both parties, could not be effectual unless some of the troops were introduced above the forts at the same time that they are below. Once above, he intended to cover their landing at Quarantine, five miles above, they coming to the river through the bayou there. Once above, the forts were cut off and his propellers intact for ascending the river to the city. And in passing the forts, if he found his ships able to cope with them, he should fight it out. Some of the captains and commanders considered it a hazardous thing to go above, as being out of the reach of supplies. To this it may be said that the steamers can pass down at the rate of twelve miles an hour. The flag-officer remarked that our ammunition is being rapidly consumed without a supply at hand, and that something must be done immediately. He believed in celerity. It was proposed by myself and assented to by the flag-officer, that three steamers should go up the river shortly after dark, under my own guidance, to break the boom."

It appears from this account, supported by the general order issued immediately after it and given a few pages further on, that Farragut had definitely determined not to await the reduction of the forts, because the bombardment so far did not indicate any probability of effectual results. It was his deliberate opinion that the loss of time and the waste of effort were entailing greater risks than would be caused by cutting adrift from his base and severing his own communications in order to strike at those of the enemy. It is commonly true that in the effort to cut the communications of an opponent one runs the risk of exposing his own; but in this case the attacking force was one pre-eminently qualified to control the one great medium of communication throughout that region—that is, the water. Also, although in surrendering the river Farragut gave up the great line of travel, he kept in view that the bayou system offered an alternative, doubtless greatly inferior, but which, nevertheless, would serve to plant above the forts, under the protection of the navy, such troops as should be deemed necessary; and that the combined efforts of army and navy could then maintain a sufficient flow of supplies until the forts fell from isolation. Finally, a fleet is not so much an army as a collection of floating fortresses, garrisoned, provisioned, and mobile. It carries its communications in its hulls, and is not in such daily dependence upon external sources as is the sister service.

In deciding, therefore, against awaiting the reduction of the forts by direct attack, and in favor of attempting the same result by striking at the interests they defended and the base on which they rested, Farragut was guided by a calculation of the comparative *material* risks and advantages of the two courses, and not mainly by consideration of the moral effect produced upon the defenders by a successful stroke, as has been surmised by Lord Wolseley. This eminent English authority attributes the success of the expedition against New Orleans to three causes. "First, the inadequate previous preparation of the naval part of the New Orleans defenses; second, the want of harmonious working between the Confederate naval and military forces; and, lastly, Farragut's clear appreciation of the moral effect he would produce by forcing his way past the defenses of Fort Jackson and Fort St. Philip, and by his appearance before New Orleans. For, after all, the forts were never captured by actual attack. . . This brilliant result is a striking instance of the due appreciation by a commander of the effect which daring achievements exert on men's minds, although, *as in this case*, those daring acts *do not actually, directly, or materially* make certain the end or surrender they may have secured." And, again, in another place: "Admiral Farragut's success was mainly due to the moral effect produced by his gallant passage of the forts. . . . He never reduced the forts, and seems to have done them but little harm." *

The moral effect produced in war upon men's minds, and through the mind upon their actions, is undeniable, and may rightly count for much in the calculations of a commander; but when it becomes the sole, or even the chief reliance, as in Bonaparte's advance into Carinthia in 1797, the spirit displayed approaches closely to that of the gambler who counts upon a successful bluff to disconcert his opponent. The serious objection to relying upon moral effect alone to overcome resistance is that moral forces do not admit of as close knowledge and measurement as do material conditions. The insight and moral strength of the enemy may be greater than you have means of knowing, and to

* Lord Wolseley in *North American Review*, vol. cxlix, pp. 32-34, 597. The italics are the author's.

assume that they are less is to fall into the dangerous error of despising your enemy. To attribute to so dubious a hope, alone, the daring act of Admiral Farragut in passing the forts and encountering the imperfectly known dangers above, is really to detract from his fame as a capable as well as gallant leader. That there were risks and accidents to be met he knew full well; that he might incur disaster he realized; that the dangers above and the power of the enemy's vessels might exceed his expectations was possible; war can not be stripped of hazard, and the anxiety of the doubtful issue is the penalty the chieftain pays for his position. But Farragut was convinced by experience and reflection that his fleet could force its passage; and he saw that once above the material probabilities were that army and navy could be combined in such a position of vantage as would isolate the forts from all relief, and so "actually, directly, and materially make certain their surrender," and secure his end of controlling the lower Mississippi. There was only one road practicable to ships to pass above, and that led openly and directly under the fire of the forts; but having passed this, they were planted across the communications as squarely as if they had made a circuit of hundreds of miles, with all the secrecy of Bonaparte in 1800 and in 1805. Are strongholds never "captured" unless by "actual attack"? Did Ulm and Mantua yield to blows or to isolation?

Such, certainly, was the opinion of the able officers who conducted the Confederate defense, and whose conduct, except in matters of detail, was approved by the searching court of inquiry that passed upon it. "In my judgment," testified General M. L. Smith, who commanded the interior line of works and was in no way responsible for the fall of Forts St. Philip and Jackson, "the forts were impregnable *so long as they were in free and open communication with the city*. This communication was not endangered while the obstruction existed. The conclusion, then, is briefly this: While the obstruction existed the city was safe; when it was swept away, as the defenses then existed, it was in the enemy's power."* General Lovell, the commander-in-chief of the military department, stated that he had made preparations to evacuate New Orleans in case the fleet passed the fort by sending out of the city several hundred thousand rations and securing transport steamers. He continued: "In determining upon the evacuation of the city I necessarily, as soon as the enemy's fleet had passed the forts, regarded the position *the same as if both their army and navy were present before the city*, making due allowance simply for the time it would take them to transport their army up. Inasmuch as their ships had passed Forts Jackson and St. Philip, *they could at once place themselves in open and uninterrupted communication with their army at points from six to twenty miles above the forts through various small water communications from the Gulf*, made more available by the extraordinary height of the river, and which, while they (we ?) were in possession of the latter, I had easily and without risk defended with launches and part of the river-defense fleet. I had also stationed Szymanski's regiment at the Quarantine for the same object. These were, however, all destroyed or capt-

* *Official Records of the War of the Rebellion*. Series I, vol. vi, p. 583.

ured by the enemy's fleet after they got possession of the river between the forts and the city." * Colonel Szymanski testified: "After the forts had been passed, it was practicable for the enemy to transport his army through the bayous and canals to New Orleans, without encountering the forts. A portion of the enemy did come that way. I have for many years owned a plantation fifteen miles below the city, and am very familiar with the whole country. I have never known the river as high as it was in 1862. Also, above English Turn (five miles below the city) there is water communication through Lake Borgne with the Gulf of Mexico by other bayous and canals of the same character." †

It is evident, therefore, that competent military men on the spot, and in full possession of all the facts, considered, as did Farragut, that with the passage of the forts by the fleet the material probabilities of success became in favor of the United States forces. The only moral effect produced was the mutiny of the half-disciplined alien troops that garrisoned the forts; and surely it will not be contended that any such wild anticipation as of that prompted Farragut's movement. The officers of the forts were trained and educated soldiers, who knew their duty and would not be crushed into submission by adverse circumstances. They would doubtless have replied, as did the commander of Fort Morgan two years later, that they looked upon the United States fleet above them as their prisoners, and they would have held out to the bitter end; but the end was certain as soon as the fleet passed above

---

* *Official Records of the War of the Rebellion.* Series I, vol. vi, p. 566.

† Ibid., p. 578.

them. They had provisions for two months; then, if not reduced by blows, they must yield to hunger.

Immediately after the conference with his captains, Farragut issued the following general order, from which it appears that, while his opinion remained unchanged as to the expediency of running by the forts, he contemplated the possibility, though not the probability, of their being subdued by the fire of the fleet, and reserved to himself freedom to act accordingly by prescribing a simple signal, which would be readily understood, and would convert the attempt to pass into a sustained and deadly effort to conquer:

"UNITED STATES FLAG-SHIP HARTFORD,
MISSISSIPPI RIVER, *April* 20, 1862.

"The flag-officer, having heard all the opinions expressed by the different commanders, is of the opinion that whatever is to be done will have to be done quickly, or we shall be again reduced to a blockading squadron, without the means of carrying on the bombardment, as we have nearly expended all the shells and fuses and material for making cartridges. He has always entertained the same opinions which are expressed by Commander Porter—that is, there are three modes of attack,* and the question is, which is the one to be adopted? His own opinion is that a combination of two should be made, viz., *the forts should be run, and when a force is once above the forts to protect the troops they should be landed at Quarantine from the Gulf side by bringing them through the bayou,* and then our forces should move

* Those three were: First, a direct naval attack upon the works; second, running by the works; third, a combined attack by army and navy.

up the river, mutually aiding each other as it can be done to advantage.

"When in the opinion of the flag-officer the propitious time has arrived, the signal will be made to weigh and advance to the conflict. If, in his opinion, at the time of arriving at the respective positions of the different divisions of the fleet we have the advantage, he will make the signal for close action, No. 8, and abide the result—conquer or be conquered—drop anchor or keep under way, as in his opinion is best.

"*Unless the signal above mentioned is made*, it will be understood that the first order of sailing will be formed after leaving Fort St. Philip, and we will proceed up the river *in accordance with the original opinion expressed.*

"The programme of the order of sailing accompanies this general order, and the commanders will hold themselves in readiness for the service as indicated.

D. G. FARRAGUT,
Flag-officer Western Gulf Blockading Squadron."

Nothing can be clearer than that the opinion expressed and maintained by the flag-officer from the beginning was the one carried out, resulting in a complete success.

The bombardment by the mortar flotilla was continued three days longer, at the end of which time the provision of bombs immediately obtainable was becoming exhausted. Enough, however, remained to sustain a very vigorous fire during the period of the passage, and as the cover of darkness was desired the delay was not without its advantages, for the waning moon grew daily less and rose an hour later each succeeding night. On the 23d notice was given to the ships that the attempt to pass would be made that night, and that, as half-past three was the hour of moon-rise, the signal, two red lights, would be hoisted at 2 A. M. During that afternoon Farragut personally visited each ship, in order to know positively that each commander understood his orders for the attack, and to see that all was in readiness.

The original intention of the flag-officer was to attack in two parallel columns, a more compact formation than one long one, less liable to straggling, and in which the heavy batteries of the larger ships would more effectually cover the lighter vessels by keeping down the fire of the enemy. In this arrangement, which remained unaltered until the 23d, the second in command, Captain Theodorus Bailey, whose divisional flag was flying in the gunboat Cayuga, would have had the right column, and the flag-officer himself the left in the Hartford. The latter was to be followed by the Brooklyn and Richmond, and upon these three heavy ships would fall the brunt of the engagement with Fort Jackson, the more powerful of the enemy's works. The right column also had its heaviest ships in the lead; the exceptional station of the Cayuga being due to some natural unwillingness on the part of other commanding officers to receive on board, as divisional commander and their own superior, an officer whose position in the fleet was simply that of captain of a single ship.* The Cayuga led, not in virtue of her

* Captain Bailey commanded the Colorado frigate, which drew too much water to cross the bar. Anxious to share in the fight, he obtained from the flag-officer the divisional appointment.

armament, but because she bore on board the commander of one column.

On the 23d Farragut, considering the narrowness of the opening in the obstructions through which the fleet must pass, decided that the risk of collision with the hulks on either side, or between the columns themselves, would be too great if he adhered to his written programme; and he accordingly gave a verbal order that the right column should weigh first, and be followed closely by the other under his own guidance. To facilitate the departure and avoid confusion, the ships of the right shifted their berth after dark to the east side of the river, anchoring in the order prescribed to them.

As some doubts had been expressed as to the actual rupture of the chains between the hulks on either side the breach, although they had evidently been dragged from their position by the efforts made on the night of the 20th, Lieutenant Caldwell was again chosen, at his own request, to make an examination of the actual conditions. This he did in the early part of the night, before the ships got under way; and it is a singular confirmation of the slackness and inefficiency that has been charged against the water service of the Confederates that he effected this duty thoroughly and without molestation. Twice he pulled above the hulks and thence allowed his boat to drift down between them, a heavy lead with sixty feet of line hanging from her bows. As this line caught on nothing it was clear that within the narrow limits of the breach no impediment to the passage of a vessel existed. By 11 P. M. Caldwell was on his return with this decisive and encouraging report.

At 2 A. M. the appointed signal was made, and at once was heard in every direction the clank-clank of the chains as the seamen hove the anchors to the bows. The strength of the current and the tenacity of the bottom in some spots made this operation longer than had been expected, and not till half-past three did the leading vessel reach the line of hulks, followed closely by the rest of her division. There is something singularly impressive in the thought of these moments of silent tension, following the active efforts of getting under way and preceding the furious strife, for whose first outburst every heart on board was waiting; and the impression is increased by the petty size of the little vessel in the lead, which thus advanced with steady beating of the engines to bear the first blast of the storm. Favored partly by her size, and yet more by the negligence of those among the enemy whose duty it was to have kept the scene alight with the numerous fire-rafts provided for that very purpose, the Cayuga passed the hulks and was well on her way up river before she was seen. "Although it was a starlight night," wrote Lieutenant Perkins, who by her commander's direction was piloting the ship, "we were not discovered until well under the forts; then they opened upon us a tremendous fire." It was the prelude to a drama of singular energy and grandeur, for the Confederates in the forts were fully on their guard, and had anticipated with unshaken courage, but with gloomy forebodings, an attack during that very night. "There will be no to-morrow for New Orleans," had said the undaunted commander of Fort Jackson the day before, "if the navy does not at once move the Louisiana to the position assigned to her," close to the obstructions. The Louisiana was a powerful ironclad battery, not quite complete when Farragut entered the river. She had been hurried down to the forts four days before the passage of the fleet, but her engines could not drive her, and the naval commander refused to take up the position, asked of him by the military authorities, below St. Philip, where he would have a cross fire with the forts, a close command of the line of obstructions, and would greatly prolong the gantlet of fire through which the fleet must run. To support the movement of the latter by drawing the fire and harassing the gunners of the enemy, Commander Porter moved up with the steamers of the mortar flotilla to easy range of the water battery under Fort Jackson, which he engaged; while the mortar schooners, as soon as the flash of the enemy's guns showed that the head of the column had been discovered, opened a furious bombardment, keeping two shells constantly in the air. Except for the annoyance of the bombs, the gunners of the forts had it much their own way until the broadsides of the Pensacola, which showed eleven heavy guns on either side, drew up abreast of them. "The Cayuga received the first fire," writes Perkins, "and the air was filled with shells and explosives which almost blinded me as I stood on the forecastle trying to see my way, for I had never been up the river before. I soon saw that the guns of the forts were all aimed for midstream, so I steered close under the walls of Fort St. Philip; and although our masts and rigging got badly shot through our hull was but little damaged." Small as she was—five hundred tons—and with the scanty top hamper of a schooner, the

Cayuga was struck forty-two times, below and aloft.

"After passing the last battery," continues Perkins, "and thinking we were clear, I looked back for some of our vessels, and my heart jumped up into my mouth when I found I could not see a single one. I thought they all must have been sunk by the forts." This seeming desertion was due to the fact that the heavy ships—the Pensacola, Mississippi, and Oneida—had been detained by the resolute manner in which the first stopped to engage Fort St. Philip. Stopping to fire, then moving slowly, then stopping again, the reiterated broadsides of this big ship, delivered at such close range that the combatants on either side exchanged oaths and jeers of defiance, beat down the fire of the exposed barbette batteries, and gave an admirable opportunity for slipping by to the light vessels, which brought up the rear of the column and were wholly unfit to contend with the forts. The Mississippi and Oneida keeping close behind the Pensacola and refusing to pass her, the Cayuga was thus separated from all her followers.

The isolation of the Cayuga was therefore caused by her anomalous position at the head of the column, a post proper only to a heavy ship. It was impossible for her petty battery of two guns to pause before the numerous pieces of the enemy; it was equally impossible for the powerful vessels following her to hasten on, leaving to the mercy of the Confederates the gunboats of the same type that succeeded them in the order. That the Cayuga was thus exposed arose from the amiable desire of the admiral to gratify Bailey's laudable wish to share in the battle, without compelling an officer of the same grade, and junior only in number, to accept a superior on his own quarter-deck in the day of battle, when the harvest of distinction is expected to repay the patient sowing of preparation. The commander of the Cayuga, who was only a lieutenant, had reconciled these conflicting claims by volunteering to carry Bailey's divisional flag. As there is no reason to suppose that Farragut deliberately intended to offer the gunboat up as a forlorn hope by drawing the first fire of the enemy, always the most deadly, and thus saving the more important vessels, the disposition of her constitutes the only serious fault in his tactical arrangements on this occasion—a fault attributable not to his judgment, but to one of those concessions to human feelings which circumstances at times extort from all men. His first intention, an advance in two columns, the heavy ships leading and closely engaging the forts with grape and canister, while the two-gun vessels slipped through between the columns, met the tactical demands of the proposed operation. The decision to abandon this order in favor of one long, thin line, because of the narrowness of the opening, can not be challenged. This formation was distinctly weaker and more liable to straggling, but nothing could be so bad as backing, collision, or stoppage at the obstructions. In such an attack, however, as in all of Farragut's battles, it seems eminently fitting that the commander of the column should lead. The occasion is one for pilotage and example; and inasmuch as the divisional commander can not control, except by example, any ship besides the one on board which he himself is, that ship should be the most powerful in his command. These conclusions may hereafter be modified by conditions of submarine warfare, though even under them it seems likely that in forcing passage into a harbor the van ship should carry the flag of the officer commanding the leading division; but under the circumstances of Farragut's day they may be accepted as representing his own convictions, first formed by the careful deliberation of a man with a genius for war, and afterward continually confirmed by his ever-ripening experience.

Left thus unsupported by the logical results of her false position, the Cayuga found herself exposed to an even greater danger than she had already run from the guns of the stationary works. "Looking ahead," says Perkins's letter, already quoted, "I saw eleven of the enemy's gunboats coming down upon us, and it seemed as if we were 'gone' sure." The vessels thus dimly seen in the darkness of the night were a heterogeneous, disorganized body, concerning which, however, very imperfect and very exaggerated particulars had reached the United States fleet. They were freely spoken of as ironclad gunboats and ironclad rams, and the Confederates had done all in their power to increase the moral effect which was attendant upon these names, then new to maritime warfare. None of them had been built with any view to war. Three only were sea-going, with the light scantling appropriate to their calling as vessels for freight and passenger traffic. Another had been a large twin-screw tugboat that began her career in Boston, and thence, shortly before the war, had been sent to the Mississippi. After the outbreak of hostilities she had been covered with an arched roof and three-quarter-inch iron; a nine-inch gun, capable only of firing directly ahead, had been mounted in her bows, and, thus equipped, she passed into notoriety as the ram Manassas. With the miserable speed of six knots, to which, however, the current of the river gave a very important addition, and with a protection scarcely stronger than the buckram armor of the stage, the Manassas, by her uncanny appearance and by the persistent trumpeting of the enemy, had obtained a very formidable reputation with the United States officers, who could get no reliable information about her.

The remainder of the force were river steamboats, whose machinery was protected with cotton, and their stems shod with one-inch iron, clamped in place by straps of the same material extending a few feet aft. Thus strengthened, it was hoped that with the sharpness of their bows and the swiftness of the current they could, notwithstanding the exceeding lightness of their structure, penetrate the hulls of the United States ships. Resolutely and vigorously handled, there can be little doubt that they might have sunk one or two of their assailants; but there is no probability that they could under all the circumstances have done more. The obscurity of the night, the swiftness of the stream, and the number of actors in the confusing drama being played between the two banks of the Mississippi, would have introduced into the always delicate fencing of the ram extraordinary difficulties, with which the inexperience of their commanders was in no degree qualified to deal. The generally steady approach, bows on, of the United States ships, presented the smallest target to their thrust and gave to the threatened vessel the utmost facilities for avoiding the collision or converting it into a glancing blow; while, as for rounding-to, to ram squarely on the beam of a ship stemming the current, the assailant, even if he displayed the remarkable nicety of judgment required, was not likely to find the necessary room.

These difficulties received illustration by the career of the Manassas that night. Her commander, Lieutenant Warley, was a former officer of the United States Navy, and he handled her with judgment and the utmost daring. Rushing nearly bows on upon the Pensacola, the thrust was wholly avoided by the quick moving of the latter's helm, which Warley characterized as beautiful; while the attempt made immediately afterward upon the Mississippi resulted in a merely glancing blow, which took a deep and long shaving out of the enemy's quarter, but did no serious damage. Not till a much later period of the action did the Manassas find an opportunity to charge squarely upon the beam of the Brooklyn. She did so across the current, striking therefore only with her own speed of six knots. But little shock was felt on board the rammed ship, and no apprehension of damage was experienced; but it was afterward found that the enemy's stem had entered between two frames, and crushed both the outer and inner planking. A few moments earlier the Brooklyn had been thrown across the current by the chances of the night. Had the ram then struck her in the same place, carrying the four knots additional velocity of the current, it is entirely possible that the mortification of the Confederate defeat would have derived some consolation from the sinking of one of Farragut's best ships. Such were the results obtained by a man of singular and resolute character, who drove his tiny vessel through the powerful broadsides of the hostile fleet, and dared afterward to follow its triumphant course up the river, in hopes of snatching another chance from the jaws of defeat.

Another example, equally daring and more successful, of the power of the ram, was given that same night by Kennon, also an ex-officer of the United States Navy; but the other ram commanders did not draw from their antecedent training and habits of thought the constancy and pride, which could carry their frail vessels into the midst of ships that had thus victoriously broken their way through the bulwarks of the Mississippi. The River-Defense Fleet, as it was called, was a separate organization, which owned no allegiance and would receive no orders from the navy; and its absurd privileges were jealously guarded by a government whose essential principle was the independence of local rights from all central authority. Captains of Mississippi River steamboats, their commanders held to the full the common American opinion that the profession of arms differs from all others in the fact that it requires no previous training, involves no special habits of thought, is characterized by no moral tone which only early education or years of custom can impart. Rejecting all suggestion and neglecting all preparation, they cherished the most inordinate confidence in the raw native valor which they were persuaded would inspire them at the critical moment; and, incredible as it would seem, some of the men who in the battle could find no other use for their boats but to run them ashore and burn them, ventured to tell Warley the night before that their mission was to show naval officers how to fight. They did not lack courage, but that military habit upon whose influence Farragut had so acutely remarked when a youth, returning in 1820 from the European station.* "Had regular naval officers," said Kennon bitterly, "instead of being kept in the mud forts on the creeks in Virginia, and in the woods of Carolina cutting timbers to build ironclads, been sent to command these vessels, even at the eleventh hour, they would have proved very formidable."

Steaming into the midst of such as these, the peril of the Cayuga, real enough, was less than it seemed; but she had to do at once with Warley's Manassas and with the Governor Moore, the vessel that Kennon commanded, and which afterward sunk the Varuna. "Three made a dash to board us," records Lieutenant Perkins, agreeing therein with the official reports of Captain Bailey and of his own commander, Lieutenant Harrison; "but a heavy charge from our eleven-inch gun settled the Governor Moore, which was one of them. A ram, the Manassas, in attempting to butt us just missed our stern, and we soon settled the third fellow's 'hash.' Just then some of our gunboats which had passed the forts came up, and then all sorts of things happened." This last expression is probably as terse and graphic a summary of a *mêlée*, which to so many is the ideal of a naval conflict, as ever was penned. "There was the wildest excitement all round. The Varuna fired a broadside into us instead of into the enemy. Another of our gunboats attacked one of the Cayuga's prizes; I shouted out, 'Don't fire into that ship, she has surrendered.'

---

* See page 62.

Three of the enemy's ships had surrendered to us before any of our vessels appeared; but when they did come up we all pitched in, and settled the eleven rebel vessels in about twenty minutes." Besides the eleven armed boats known to have been above, there were several unarmed tugs and other steamers, some of which probably shared in this wild confusion. One at least came into conflict with the Hartford.

The second column, led by the flag-ship, was promptly away and after the first; following, indeed, so closely that the head of the one lapped the rear of the other. The Brooklyn and Richmond, close behind the Hartford, formed with her a powerful "body of battle," to use the strong French expression for the center of a fleet. Though called sloops-of-war, the tonnage and batteries of these ships were superior to those of the medium ships-of-the-line of the beginning of this century, with which Nelson fought his celebrated battles. As the flag-ship reached the hulks the night, which, though very dark, was fairly clear, had become obscured by the dense clouds of smoke that an almost breathless atmosphere suffered to settle down upon the water. Only twenty minutes had elapsed since the forts opened upon the Cayuga, when Farragut's flag entered the battle. Soon after passing the obstructions, and when about to sheer in toward Fort Jackson, upon which was to be concentrated her own battery and that of her two formidable followers, a fire-raft was observed coming down the river in such a way as to make contact probable if the course were not changed. Heading across the river, and edged gradually over by the raft continuing to work toward her, the ship took the ground a little above Fort

St. Philip, but still under its batteries. While in this dangerous position, the raft, whose movements proved to be controlled not by the current but by a small tugboat, was pushed against her port quarter. The flames caught the side of the ship, spread swiftly along it, leaped into the rigging and blazed up toward the tops. The danger was imminent, and appeared even more so than it was; for the body of heat, though great, was scarcely sufficient to account for such a rapid spread of the flames, which was probably due mainly to the paint. The thoroughly organized fire department soon succeeded in quenching the conflagration, its source being removed by training some of the after-guns upon the daring pygmy, which with such reckless courage had well-nigh destroyed the commander-in-chief of her enemy's fleet. The tug received a shot in her boilers and sunk. The Hartford backed clear, but in so doing fell off broadside to the stream, thereby affording another chance to the hostile rams, had there been one prepared to dare the hazard. Watson, the flag-lieutenant, remarks that the flag-officer stood during this critical period giving his orders and watching the ship slowly turn, referring occasionally to a little compass which was attached to his watch-chain. During most of the engagement, however, he was forward observing the conflict.

The Brooklyn and Richmond, with the Sciota and the Iroquois, which followed immediately after them, fought their way through with more or less of adventure, but successfully reached the river above the forts. It is to be observed, however, that these, as well as the Hartford, suffered from the embarrassment of the smoke, which had inconvenienced the ships of the first column to a much less degree. This was to be expected, and doubtless contributed to the greater loss which they suffered, by delaying their progress and giving uncertainty to their aim; the result of the latter being naturally to intensify the action of the hostile gunners. Four gunboats brought up the rear of the column, of which but one got through, and she with a loss greater than any vessel of her class. The three last failed to pass. Blinded by smoke and further delayed by the tendency to open out, which is observable in all long columns, they came under the fire of the forts at a time when, the larger vessels having passed, they were no longer covered or supported by their fire, and when day was about to break. The Itasca, commanded by the gallant Caldwell, who had so nobly broken through the obstructions, opposing only her puny battery to the concentrated wrath of the forts, was knocked about by them at will, received a shot through her boiler and drifted down the river out of action. The Winona likewise encountered almost alone, or perhaps in company with the Itasca, the fire of the enemy. After nearly running ashore in the smoke, daylight surprised her while still under fire below the works; and her commander very properly decided not to risk the total destruction and possible capture of his vessel for the sake of adding her insignificant force to that above. Admirably as the gunboats were officered, perhaps their most useful service on this night was to demonstrate again the advantage of big ships, as of big battalions.

Thirteen out of his seventeen vessels having rallied around his flag above the forts, and the three below being of the least efficient type, the flag-officer could congratulate himself upon a complete victory, won with but little loss. One vessel only was sacrificed, and she to that inconsiderate ardor which in so many cases of pursuit leads men, without any necessity, out of reach of support. The Varuna, the fifth in the order, and the only merchant-built vessel in the fleet, after clearing the forts had steamed rapidly through the Confederate flotilla, firing right and left, but not stopping. She soon passed above it, and getting sight of a small steamer heading for New Orleans, sped away after her. Kennon, in the Governor Moore, happened to have noticed this movement; and, finding by the rapid accessions to the number of his enemies that he was likely to be soon overwhelmed, he determined to follow this one which, whatever her strength, he might tackle alone. Stealing out of the *mêlée* he started up the river, hoisting lights similar to those he had observed the enemy's ships to carry. Deceived by this ruse, the Varuna at the first paid no attention to her pursuer, some distance behind whom followed one of the River-Defense boats, the Stonewall Jackson. When Kennon at last opened fire, the Varuna, having by then run down her steam in her headlong speed, was being rapidly overtaken. The second shot from the Moore raked the Varuna's deck, killing and wounding twelve men. The Union vessel's helm was then put hard-a-port, swinging her broadside to bear upon her approaching foe, who was naturally expected to imitate the movement, opposing side to side to avoid being raked. Instead of so doing Kennon kept straight on, and, while receiving a deadly raking fire from his antagonist's battery, which struck down many of his men, he succeeded in driving the sharp stem of the Moore through the side of the Varuna. A few moments after the Stonewall Jackson coming up also rammed the disabled enemy, whose commander then drove her ashore on the east side of the river, where she sank. By this time the corvette Oneida had made out the state of the case. Steaming rapidly ahead, she overhauled the Confederate vessels; which, finding they could not escape, ran ashore, the Jackson on the west bank, the Moore on the east, and in those positions they were surrendered.

Farragut had undertaken this daring exploit with the expectation that, after passing the forts, he could obtain the co-operation of the army, and that the action of the two services, combined in mutual support, would suffice to force the way to New Orleans. The occupation of the land by the army, and of the water by the navy, interposing by the nature of their operations between the city and the forts, would effectually isolate the latter. In accordance with this plan he at once sent Captain Boggs, of the Varuna, through the Quarantine Bayou with messages to Commander Porter and General Butler. The latter was notified that the way was now clear to land his troops through the bayou, in accordance with the previous arrangements, and that gunboats would be left there to protect them against those of the enemy, of which three or four were seen to be still at the forts. Boggs passed successfully through the country and streams which a day before had been in quiet possession of the enemy, though it took him twenty-six hours to do so; but General Butler, who from a transport below had witnessed the success of the fleet, had waited for no further tidings.

Hurrying back to his troops, he collected them at Sable Island, twelve miles in rear of Fort St. Philip, whence they were transported and landed at a point on the river five miles above the work, where the Kineo and Wissahickon awaited them.

During the remainder of the 24th the fleet stayed at anchor off the Quarantine station, to repose the crews after the excessive labor and excitement of the previous night. Early the next morning all got under way except the two gunboats left to support Butler's troops, and moved up stream; but slowly, owing to the indifferent speed of some and to want of knowledge of the river. At half-past ten they reached English Turn, five miles below the city; the point where the British forces had in 1815 been so disastrously repelled in their assault upon the earthworks held by Jackson's riflemen. The Confederates had fortified and armed the same lines on both sides of the Mississippi, as part of the interior system of defenses to New Orleans; the exterior line being constituted by Forts Jackson and St. Philip, together with several smaller works at different points, commanding the numerous subsidiary approaches through the Mississippi delta. The interior lines at English Turn, known as the Chalmette and McGehee batteries, were, however, intended only to check an approach of troops from down the river. Their general direction was perpendicular to the stream; and along its banks there ran only a short work on either side to protect the main entrenchments from an enfilading fire by light vessels, which might, in company with an invading army, have managed to turn the lower forts by passing through the bayous. These river batteries, mounting respectively nine and five guns, were powerless to resist the ships that had successfully passed the main defenses of the city. After a few shots, fired rather for the honor of the flag than in any hope of successful result, the guns were forsaken; and both lines of entrenchments, being turned and taken in the rear, were abandoned.

Meanwhile, in New Orleans a scene of fearful confusion was growing hourly more frenzied. Whatever the fears of the military commanders as to the result of the attack upon the forts, they had very properly concealed them from the inhabitants; and these, swayed by the boastful temper common to mobs, had been readily led to despise the efforts of the enemy and to trust implicitly in the power of their defenses. General Lovell, commanding the department, had gone down to the forts the evening before the attack, and was still there when the United States fleet was breaking its way through; he was, in fact, on board the little steamer, the pursuit of which lured the Varuna into the isolation where she met her fate. The news of the successful forcing of the exterior and principal defenses thus reached the city soon after it was effected; and at the same time Lovell, satisfied from the first that if the forts were passed the town was lost, prepared at once to evacuate it, removing all the Government property. This in itself was a service of great difficulty. New Orleans is almost surrounded by water or marsh; the only exit was to the northward by a narrow strip of dry land, not over three quarters of a mile wide, along the river bank, by which passed the railroad to Jackson, in the State of Mississippi. As has already been said, Lovell had by this road been quietly removing army rations for some time, but had abstained from trying to carry off any noticeable articles by which his apprehensions would be betrayed to the populace. The latter, roused from its slumber of security with such appalling suddenness, gave way to an outburst of panic and fury; which was the less controllable because so very large a proportion of the better and stronger element among the men had gone forth to swell the ranks of the Confederate army. As in a revolution in a South American city, the street doors were closed by the tradesmen upon the property in their stores; but without began a scene of mad destruction, which has since been forcibly portrayed by one, then but a lad of fourteen years, who witnessed the sight.

Far down the stream, and throughout their ascent, the ships were passing through the wreckage thus made. Cotton bales, cotton-laden ships and steamers on fire, and working implements of every kind such as are used in ship-yards, were continually encountered. On the piers of the levees, where were huge piles of hogsheads of sugar and molasses, a mob, composed of the scum of the city, men and women, broke and smashed without restraint. Toward noon of the 25th, as the fleet drew round the bend where the Crescent City first appears in sight, the confusion and destruction were at their height. "The levee of New Orleans," says Farragut in his report, "was one scene of desolation. Ships, steamers, cotton, coal, etc., were all in one common blaze, and our ingenuity was much taxed to avoid the floating conflagration. The destruction of property was awful." Upon this pandemonium, in which the fierce glare of burning property lit up the wild passions and gestures of an infuriated people, the windows of heaven were opened and a drenching rain poured down in torrents. The impression produced by the ships as they came in sight around the bend has been graphically described by the boy before mentioned, who has since become so well-known as an author—Mr. George W. Cable. "I see the ships now, as they come slowly round Slaughter House Point into full view, silent, grim, and terrible; black with men, heavy with deadly portent, the long-banished Stars and Stripes flying against the frowning sky. Oh! for the Mississippi! for the Mississippi!" (an iron-clad vessel nearly completed, upon which great hopes had been based by the Confederates). "Just then she came down. But how? Drifting helplessly, a mass of flames.

"The crowds on the levee howled and screamed with rage. The swarming decks answered never a word; but one old tar on the Hartford, standing lanyard in hand beside a great pivot gun, so plain to view that you could see him smile, silently patted its big black breech and blandly grinned. And now the rain came down in torrents."

That same morning, as though with the purpose of embarrassing the victor whom he could not oppose, the Mayor of New Orleans had ordered the State flag of Louisiana to be hoisted upon the City Hall. His secretary, who was charged with this office, waited to fulfill it until the cannonade at English Turn had ceased, and it was evident the fleet had passed the last flimsy barrier and would within an hour appear before the city. The flag was then run up; and the Mayor had the satisfaction of creating a position of very unnecessary embarrassment for all parties by his useless bravado.

To Captain Bailey, the second in command, who

had so gallantly led both in the first assault and in the attack at Chalmette, was assigned the honor of being the first to land in the conquered city and to demand its surrender. It was no barren honor, but a service of very sensible personal danger to which he was thus called. General Lovell having to devote his attention solely to his military duties, the city which had so long been under martial law was escaping out of the hands of the civil authorities and fast lapsing into anarchy. Between one and two in the afternoon Bailey landed, accompanied by Perkins, the first lieutenant of the Cayuga; who, having shared his former perils, was permitted to accompany him in this one also. "We took just a boat and a boat's crew," writes Perkins, "with a flag of truce, and started off. When we reached the wharf there were no officials to be seen; no one received us, although the whole city was watching our movements, and the levee was crowded in spite of a heavy rainstorm. Among the crowd were many women and children, and the women were shaking rebel flags and being rude and noisy. They were all shouting and hooting as we stepped on shore. . . . As we advanced the mob followed us in a very excited state. They gave three cheers for Jeff Davis and Beauregard and three groans for Lincoln. Then they began to throw things at us, and shout 'Hang them!' 'Hang them!' We both thought we were in a bad fix, but there was nothing for us to do but just to go on." Mr. Cable has given his description of the same scene: "About one or two in the afternoon, I being in the store with but one door ajar, came a roar of shoutings and imprecations and crowding feet down Common Street. 'Hurrah for Jeff Davis!'

'Shoot them!' 'Kill them!' 'Hang them!' I locked the door of the store on the outside and ran to the front of the mob, bawling with the rest, 'Hurrah for Jeff Davis!' About every third man had a weapon out. Two officers of the United States navy were walking abreast, unguarded and alone, not looking to the right or left, never frowning, never flinching, while the mob screamed in their ears, shook cocked pistols in their faces, cursed, crowded, and gnashed upon them. So through those gates of death those two men walked to the City Hall to demand the town's surrender. It was one of the bravest deeds I ever saw done."

Farragut's demand, made through Bailey, was that the flag of Louisiana should be hauled down from the City Hall, and that of the United States hoisted over the buildings which were its property, namely, the Custom House, Post Office, and Mint. This the Mayor refused to do; and, as Farragut had no force with which to occupy the city, it became a somewhat difficult question to carry on an argument with the authorities of a town protected by the presence of so many women and children. The situation was for three days exceedingly critical, from the temper and character of the mob and from the obstinacy and powerlessness of the officials. It was doubtless as much as the life of any citizen of the place was worth to comply with the admiral's demands. On the other hand, while there could be no difficulty in hoisting the United States flag, there would be much in protecting it from insult with the means at the flag-officer's disposal; for to open fire upon a place where there were so many helpless creatures, innocent of any greater offense than behaving like a set of spoiled children, was a course that could not be contemplated unless in the last necessity, and it was undesirable to provoke acts which might lead to any such step. The United States officers who were necessarily sent to communicate with the authorities did so, in the opinion of the authorities themselves, at the peril of their lives from a mob which no one on shore could control. On the 28th of April, however, Forts Jackson and St. Philip surrendered to Commander Porter in consequence of a mutiny in their garrisons, which refused to fight any longer, saying further resistance was useless; and the following day Farragut sent ashore a body of two hundred and fifty marines with two howitzers manned by seamen from the Hartford, the whole under the command of the fleet-captain, Captain Henry H. Bell. The force was formally drawn up before the City Hall, the howitzers pointing up and down the street, which was thronged with people. Fearing still that some rash person in the crowd might dare to fire upon the men who were hauling down the flag, the Mayor took his stand before one of the howitzers; a sufficient intimation to the mob that were murder done he would be the first victim to fall in expiation. The United States flag was then hoisted over the Custom House, and left flying under the protection of a guard of marines.

Thus was timely and satisfactorily completed an act, by which Farragut signalized and sealed the fact that the conquest of New Orleans and of its defenses, from the original conception of the enterprise to its complete fulfillment by the customary tokens of submission and taking possession, was wholly the work of the United States Navy; of which he, by his magnificent successes, became the representative figure. It was a triumph won over formidable difficulties by a mobile force, skillfully directed and gallantly fought. By superior promptitude and a correct appreciation of the true strategic objective had been reduced to powerlessness obstacles not to be overcome by direct assault, except by a loss of time which would have allowed the enemy to complete preparations possibly fatal to the whole undertaking. Forts Jackson and St. Philip, which the fleet could not have reduced by direct attack, fell by the severance of their communications.

It is not to be questioned that the moral effect of the passage of the forts, succeeded, as it was, by the immediate fall of the great city of the Mississippi, was very great; but it was not upon the forts themselves, nor in the unexpected mutiny of the garrison, that that effect was chiefly manifested. Great as was the crime of the men, they showed by their act a correct appreciation of those results to the forts, from the passage of the fleet, which some have sought to ignore—results physical, undeniable, fatal. It was not moral effect, but indisputable reasoning which sapped the further resistance of men—brave till then—to whom were wanting the habit of discipline and the appreciation of the far-reaching effects upon the fortunes of a campaign produced by a prolonged, though hopeless, resistance. They saw that the fate of the forts was sealed, and beyond that they recognized no duties and no advantages. On the scene of his exploit Farragut reaped the material fruits of the celerity in which he believed; and which he had reluctantly for a space postponed, at the bidding of superior authority, in order to try the effect of slower methods. These being exhausted, he owed to the promptness of his decision and action that the Louisiana, on whose repairs men were working night and day, did not take the advantageous position indicated to her by the officers of the forts; and that the Mississippi, the ironclad upon which not only the designers, but naval officers, founded extravagant hopes, was neither completed nor towed away, but burned where she lay. The flaming mass, as it drifted hopelessly by the Hartford, was a striking symbol of resistance crushed—of ascendency established over the mighty river whose name it bore; but it was a symbol not of moral, but of physical victory.

It was elsewhere, far and wide, that were felt the moral effects which echoed the sudden, unexpected crash with which the lower Mississippi fell—through the length and breadth of the South and in the cabinets of foreign statesmen, who had believed too readily, as did their officers on the spot, that the barrier was not to be passed—that the Queen City of the Confederacy was impregnable to attack from the sea. Whatever may have been the actual purposes of that mysterious and undecided personage, Napoleon III, the effect of military events, whether on sea or shore, upon the question of interference by foreign powers is sufficiently evident from the private correspondence which, a few months after New Orleans, passed between Lords Palmerston and Russell, then the leading members of the British Cabinet.* Fortunately for the cause of the United States, France

* See Walpole's *Life of Lord John Russell*, vol. ii, pp. 349–351.

and Great Britain were not of a mind to combine their action at the propitious moment; and the moral effect of the victory at New Orleans was like a cold plunge bath to the French emperor, at the time when he was hesitating whether to act alone. It produced upon him even more impression than upon the British Government; because his ambitions for French control and for the extension of the Latin races on the American continent were especially directed toward Louisiana, the former colony of France, and toward its neighbors, Texas and Mexico.

The sympathies, however, of the classes from whom were chiefly drawn the cabinets of the two great naval States were overwhelmingly with the South; and the expressions alike of the emperor and of his principal confidants at this time were designedly allowed to transpire, both to the Southern commissioners and to the British Government. On the very day that Porter's mortar schooners opened on Fort Jackson, Louis Napoleon unbosomed himself to a member of the British Parliament, who visited him as an avowed partisan of the Confederate cause. He said that while he desired to preserve a strict neutrality, he could not consent that his people should continue to suffer from the acts of the Federal Government. He thought the best course would be to make a friendly appeal to it, either alone or concurrently with England, to open the ports; but to accompany the appeal with a proper demonstration of force upon our coasts, and, should the appeal seem likely to be ineffectual, to back it by a declaration of his purpose not to respect the blockade. The taking of New Orleans, which he did not then anticipate, might render it inexpedient to act; that he would not decide at once, but would wait some days for further intelligence.* Similar semi-official assurances came from different persons about the emperor; and the members of the Cabinet, with a single exception, showed little reserve in their favorable expressions toward the Confederacy.

A few weeks later Mr. Slidell had a conversation with M. Billault, the minister *sans portefeuille*, one of the most conservative and cautious men in the Cabinet, who represented the Government in the Chambers upon all subjects connected with foreign affairs. Slidell read a note which he had received from Sir Charles Wood, a leading Southern sympathizer in England, denying that the British Government was unwilling to act in American affairs—a denial to which some color is given by the correspondence of Palmerston and Russell before mentioned. In answer, M. Billault declared that the French Cabinet, with the possible exception of M. Thouvénel, had been unanimously in favor of the South, and added that if New Orleans had not fallen its recognition would not have been much longer delayed; but, even after that disaster, if decided successes were obtained in Virginia and Tennessee, or the enemy were held at bay for a month or two, the same result would follow. After an interview with M. Thouvénel, about the same time, Slidell reported that, though that minister did not directly say so, his manner gave fair reason to infer that if New Orleans had not been taken, and no very serious reverses were suffered in Virginia and Tennessee, recognition would very soon have been declared.†

---

* *North American Review*, vol. cxxix, p. 347.

† Ibid., vol. cxxix, p. 348.

In its moral effect, therefore, the fall of the river forts and of New Orleans, though not absolutely and finally decisive of the question of foreign intervention, corresponded to one of those telling blows, by which a general threatened by two foes meets and strikes down one before the other comes up. Such a blow may be said to decide a campaign; not because no chance is left the enemy to redeem his misfortune, but because without the first success the weaker party would have been overwhelmed by the junction of his two opponents. The heart-rending disasters to our armies during the following summer does but emphasize the immense value to the Union cause of the moral effect produced by Farragut's victory. Those disasters, as it was, prompted the leaders of the British ministry to exchange confidences in which they agreed on the expediency of mediation. They did not carry all their colleagues with them; but who can estimate the effect, when the scales were thus balancing, if the navy had been driven out of the Mississippi as the army was from Virginia?

CHAPTER VIII.

THE FIRST ADVANCE ON VICKSBURG.

1862.

The purpose of the Navy Department, as expressed in the original orders to Farragut, had been to send his squadron up the river immediately after the capture of New Orleans. The words were: "If the Mississippi expedition from Cairo shall not have descended the river, you will take advantage of the panic to push a strong force up the river to take all their defenses in the rear." When New Orleans fell, the Cairo expedition, more commonly known as the Mississippi flotilla, so far from having descended the river to the neighborhood of New Orleans, was still detained before Fort Pillow, one of the outlying defenses of Memphis, forty miles above the latter city and over eight hundred from New Orleans. It was not until the end of May that the evacuation of Corinth by the Confederates made Memphis untenable, leading to the abandonment of the forts on the 4th of June and the surrender of the city on the following day. It became therefore incumbent upon Farragut, after turning over the command of New Orleans to Butler on the 1st of May, to go up the river as soon as he possibly could.

Although the flag-officer seems to have acquiesced

in this programme in the beginning, it was probably with the expectation that the advance, up river and against the current, required of his heavy-draught and slow-moving ships would not be very far; that the Cairo expedition, which at the date of the orders quoted, January 20th, had not begun to move, would, from the character of the vessels composing it, many being ironclad, and from the advantage of the current, have progressed very far by the time he had taken New Orleans. Moreover, at that date the upper river flotilla was still a branch of the army, and its prospective movements were to be in combination with, and a part of, a great military enterprise, securing control both of the stream and of the land; whereas Farragut's was a purely naval operation, to which the army contributed only a force sufficient to hold the points which were first reduced by the fleet.

Under the actual conditions, the proposed ascent of the river bore a very different aspect to the commanding naval officer on the spot from that which presented itself to the fond imaginations of the officials in Washington. The question now was not one of fighting batteries, for there was no reason as yet to expect anything heavier than the fleet had already overcome with ease; it was the far more difficult matter of communications, in the broadest scope of the word, to be maintained over a long, narrow, tortuous, and very difficult road, passing in many places close under the guns of the enemy. "As I stated in my last dispatch," wrote Farragut to the department after his first visit to Vicksburg, "the dangers and difficulties of the river have proved to us, since we first entered it, much greater impediments to our progress, and more destructive to our vessels, than the enemy's shot. Between getting aground, derangement of the machinery, and want of coal, the delays in getting up the river are great." To take the defenses in the rear, and in their then state to drive the enemy out of them, was one thing; but to hold the abandoned positions against the return of the defenders, after the fleet had passed on, required an adequate force which Butler's army, calculated by McClellan for a much narrower sphere, could not afford. Coal and supply ships, therefore, must either run the gantlet for the four hundred miles which separated Vicksburg from New Orleans, or be accompanied always by armed vessels. The former alternative was incompatible with the necessary security, and for the latter the numbers of the fleet were utterly inadequate. In fact, to maintain the proposed operations, there would be needed so many ships to guard the communications that there would be none left for the operations to which they led.

It must also be observed that not only was this line of communications four times as long as that which led from the sea to New Orleans, and of far more difficult pilotage, but that the natural character of the enemy's positions upon it was essentially different. They were as yet undeveloped by art; but by nature they were high and commanding bluffs, having secure land communications with an extensive enemy's country in their rear over which our troops exercised no control whatever—where they had not even been seen. To speak of "taking them in the rear" was to beg the question—to assume that their front was then, as in June, 1863, toward an enemy investing them on the land side. New

Orleans and the region below, including its defenses and the communications therewith, were low-lying and intersected with numerous water-courses; over such a navy naturally exercises a preponderating control. Above New Orleans the low delta of the Mississippi extends, indeed, on the west bank as far as the Red River, if it may not be said to reach to Vicksburg and beyond; but on the east bank it ceases one hundred and fifty miles from the city. From thence to Vicksburg, a distance of two hundred and fifty miles, the stream is bordered by a series of bluffs backing on a firm country of moderate elevation. Such positions are not to be reduced from the water alone. On the contrary, if the water be a narrow strip swept by their guns, they command it; while, from the extent of country in their rear, they are not susceptible of isolation by fleets above and below, as were Forts Jackson and St. Philip.

This series of bluffs became, therefore, the line upon which the Confederates based their control of the Mississippi and maintained their vital communications with Texas and the Red River region. It could be reduced only by a military force; and to think of subduing it by a fleet taking advantage of the panic following the fall of New Orleans, was truly to rely upon moral effect without adequate physical force to support it. It is due to the Navy Department to say that they expected the army from the North to advance more rapidly than it did; but, without seeking to assign the blame, the utterly useless penetration of the United States fleet four hundred miles into the heart of the enemy's country and its subsequent mortifying withdrawal, when contrasted with the brilliant success resulting from Farragut's dash by the forts, afford a very useful lesson in the adaptation of means to ends and the selection of a definite objective, upon compassing which something happens. The object of the United States Government being to control the lower Mississippi, that was effected by means of isolating its defenses, which then fell. When the further object was sought of controlling the course of the stream above, the mere perambulation of a body of ships effected nothing, because it aimed at nothing in particular, and could have no effect upon the decisive points.

Of all these considerations Farragut was fully sensible; and, while he obeyed his orders, he showed in his dispatches to the Department, and in private letters of the same period, how much against his judgment were operations conceived on such erroneous military principles and undertaken with such inadequate force. The Department was forward to press him on, and as early as the 17th of May sent a dispatch intimating that he had forgotten his orders on the subject; and he was urged and required to open up the Mississippi to Flag-officer Davis's command (the Mississippi flotilla), then still above Memphis. This and other letters of the same date must have been peculiarly exasperating; for they were received early in June, when he had been up the river as far as Vicksburg and satisfied himself that without an adequate force of troops nothing could be accomplished. "The Department," he replies, "seems to have considered my fleet as having escaped all injury, and that when they arrived off New Orleans they were in condition to be pushed up the river. This was not the case; but, the moment the vessels could be gotten ready, the gunboats were all sent up under the command of Commander S. P. Lee, with directions to proceed to Vicksburg, take that place, and cut the railroad. . . . From all I could hear it was not considered proper, even with pilots, to risk the ships beyond Natchez. . . . By the time Commander Lee arrived at Vicksburg (May 18th) he was satisfied that the force of the enemy was too great for him to venture to take the town, or even to pass it. The land in the rear of Vicksburg is about two hundred feet high, on which are placed some eight and ten inch columbiads, which are perfectly secure from our fire. . . . I determined to get the heavy ships up there if possible, which I did a day or two after. General Williams arrived in the mean time with fifteen hundred men, when I proposed to him, if he could carry the battery on the hill, I would attack the town. He made a careful reconnaissance, and returned to me in the afternoon, when I had all the (naval) commanders assembled. He reported that it would be impossible for him to land, and that he saw no chance of doing anything with the place so long as the enemy were in such force, having at their command thirty thousand men within one hour by railroad. A large majority of the commanders concurred with him in the opinion."

Writing to his home about this council, in which, contrary to his independent decision when below Fort Jackson, he yielded to the advice of his captains, he said: "I did not pass Vicksburg; not because it was too strongly fortified; not because we could not have passed it easily enough, *but we would have been cut off from our supplies of coal and provisions.* We would have been placed between two enemies (Vicksburg and Memphis), and so the captains advised me not to do it. I was very sick at the time, and yielded to their advice, *which I think was good;* but I doubt if I would have taken it had I been well." Here is seen, transpiring vividly enough, the uncertainty and indecision arising from the conflict between the orders of the Department and his own sounder judgment. He would fain obey; yet no orders could override, though they might cruelly embarrass, the responsibility of the officer in command on the spot. "Fighting is nothing," he adds, "to the evils of the river—getting on shore, running foul of one another, losing anchors, etc." "The army," he resumes in his dispatch to the Department, "had been sent up early with a few days' rations, and I was compelled to supply them from the squadron, thereby reducing our own supplies, which were barely sufficient to bring the ships back to New Orleans, making allowance for probable delays. The river was now beginning to fall, and I apprehended great difficulty in getting down should I delay much longer. In the mean time coal vessels had been towed up the river just above Natchez (a hundred miles below Vicksburg), which vessels I was obliged to bring down and keep in company with the vessels of war, for fear of their being captured by the guerrilla bands which appear to infest almost the entire banks of the river wherever there are rapids and bluffs."

Such were some of the difficulties being experienced when the Assistant-Secretary of the Navy was writing: "The *only* anxiety *we* feel is to know if you have followed up your instructions and pushed a strong force up the river to meet the Western flotilla." "I had no conception," replied Farragut, "that the Department ever contemplated that the ships of this squadron were to attempt to go to Memphis, above which the Western flotilla then was; nor did I believe it was practicable for them to do so, unless under the most favorable circumstances, in time of peace, when their supplies could be obtained along the river. The gunboats are nearly all so damaged that they are certainly not in condition to contend with ironclad rams coming down upon them with the current. . . . We consider the advantage entirely in favor of the vessel that has the current added to her velocity." In conclusion he adds: "I arrived in New Orleans with five or six days' provisions and one anchor, and am now trying to procure others. As soon as provisions and anchors are obtained we will take our departure for up the river, and endeavor to carry out, as far as practicable, the orders conveyed in your different dispatches." Writing home, he expressed himself more freely and unmistakably: "They will keep us in this river until the vessels break down and all the little reputation we have made has evaporated. The Government appears to think that we can do anything. They expect me to navigate the Mississippi nine hundred miles in the face of batteries, ironclad rams, etc., and yet with all the ironclad vessels they have North they could not get to Norfolk or Richmond. . . . Well, I will do my duty to the best of my ability, and let the rest take care of itself. . . . They can not deprive me and my officers of the historical fact that we took New Orleans. Now they expect impossibilities."

Enough has been quoted to show that Farragut was in no way responsible for, nor approved of, the ill-timed tenacity with which the Government held to its original plan, when the conditions had turned out entirely different from those at first expected. The Secretary of the Navy at a later date endeavored to throw the blame of failure entirely upon the War Department, which was either unwilling or unable to support the naval movement with adequate troops. It is not necessary, in a life of the admiral, to attempt to decide upon the degree of remissness, if any, shown by the military service, nor upon whose shoulders it falls. It is sufficient to point out that the Navy Department required of Farragut to go up to meet the Western flotilla when it was near nine hundred miles from the mouth of the Mississippi, for no better reason, apparently, than that it had determined upon the junction at a time when it supposed it would be effected much lower down. In so doing it left nothing to the judgment of the officer commanding on the spot. "I think," said Farragut quietly, "that more should have been left to my discretion; but I hope for the best, and pray God to protect our poor sailors from harm." His own opinion was that Mobile should be the next point attacked. The difficulties there were not so great as those encountered at the Mississippi forts; and his success at the latter might not improbably have considerable moral effect upon the other works, whose position had some strong features of resemblance to those already subdued, and which were not yet in the strong state of defense which they afterward reached. The blockade of the coast was part of his charge; and in no way did he think it could be so thoroughly maintained as by occupying the harbors themselves, or their entrances.

In obedience to his peremptory orders Farragut again started up the river, with the apprehension that if he once got above Vicksburg he would not be able to return before the next spring rise; for the season of lowest water in the Mississippi was now at hand. The Hartford did run ashore on the way up, and remained hard and fast for the better part of twenty-four hours. "It is a sad thing to think of having your ship on a mud bank, five hundred miles from the natural element of a sailor," wrote the flag-officer; "but I knew that I had done all I could to prevent her being up the river so high, and was commanded to go." She had to take out her coal and shot, and had even removed two guns before she floated.

On the 18th of June the squadron was assembled just below Vicksburg, having in company also seventeen schooners of the mortar flotilla, still under Porter's command. These were placed as rapidly as possible in suitable positions on the two sides of the river, opened fire on the 26th, and continued it through the 27th. Upon the evening of the latter day Porter notified the flag-officer that he was ready to cover, by a steady bombardment, the intended passage of the fleet before the batteries.

Vicksburg is situated on the first high land met on the east bank of the Mississippi after leaving Memphis, from which it is four hundred miles distant. The position was one of peculiar strength and importance for commanding the navigation of the river. Not only was it exceptionally lofty, and on one flank of that series of bluffs which has before

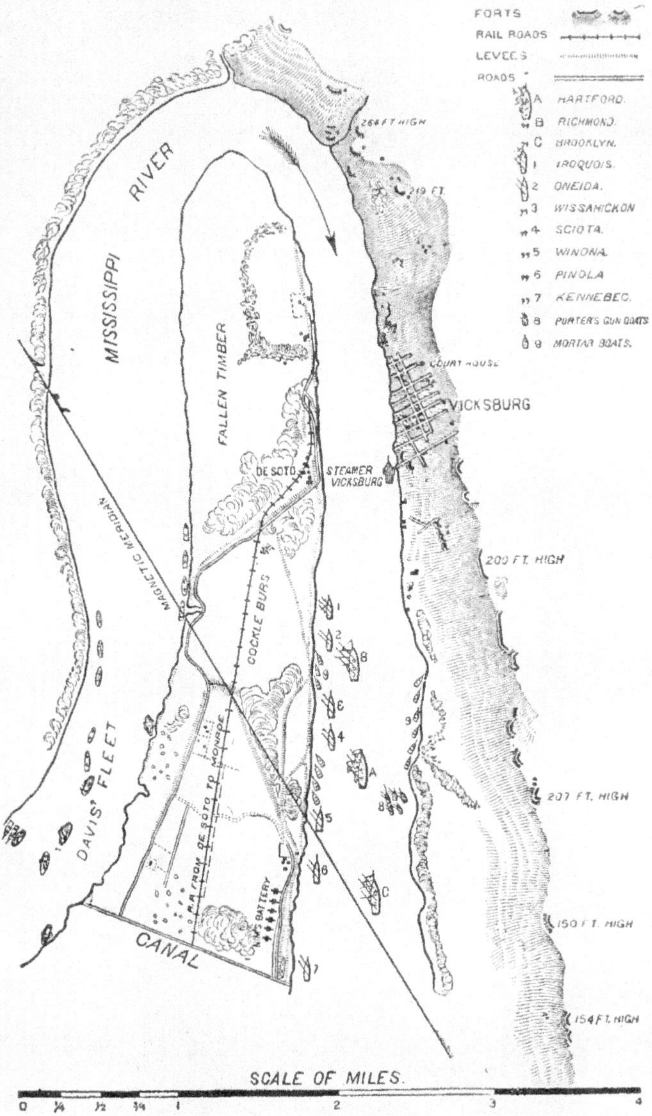

PASSAGE OF VICKSBURG BATTERIES, JUNE 28, 1862.
ORDER OF ATTACK.

been mentioned as constituting the line upon which the Confederate grip of the stream was based, but the tortuous character of the channel gave particular facilities for an enfilading fire on vessels both before and after they came abreast the works. They were thus exposed to a longer and more dangerous cannonade than is the case where the stream flows straight past the front of a battery. The channel has now changed; but in 1862 the river, which from Memphis had pursued its winding course through an alluvial country, made when abreast of Vicksburg a sharp turn to the northeast, as though determined to reach the bluffs but four miles distant. As it neared them it swung round with a sharp turn to the southwest, parallel to its recent direction, flowing for the most part close to the foot of the hills. Between the two reaches, and formed by them, immediately opposite the town, there was a low tongue of land, or promontory, four miles long and less than one wide. The squadron, being below, had to steam up through the lower reach against the current, make the sharp turn at the bend, and then pass through the upper reach. In the bend it was followed by a fire from the highest part of the bluffs, to which it could make no reply.

At 2 A. M. of June 28th the signal was given, and at three the squadron was under way—eleven vessels, of which three were the heavy ships Hartford, Richmond, and Brooklyn; two, the corvettes Iroquois and Oneida; and six gunboats. At four, the ships in their slow progress, stemming the current, had passed the mortar schooners; and the latter then opened fire, as did the steamers connected with them, which were not to attempt the passage. Owing to a misunderstanding, the three vessels which formed the rear of the column, the Brooklyn and two gunboats, did not get by. The others, at 6 A. M., anchored above Vicksburg. Though exposed much of the time to a raking fire, to which they were not able to reply, the vessels suffered less than would have been expected, owing to the enemy falling into the common mistake of giving too much elevation to his guns. Having thus accomplished his instructions, Farragut reported coldly to the Department that, in obedience to the orders "and the command of the President, I proceeded up to Vicksburg with the Brooklyn, Richmond, and Hartford, with the determination to carry out my instructions to the best of my ability. . . . The Department will perceive from this report that the forts can be *passed*, and *we have done it*, and can *do it again as often as may be required* of us. It will not, however, be an easy matter for us to do more than silence the batteries for a time, as long as the enemy has a large force behind the hills to prevent our landing and holding the place." "I am satisfied," he says again, "it is not possible to take Vicksburg without an army of twelve or fifteen thousand men. General Van Dorn's division (Confederate) is here, and lies safely behind the hills. The water is too low for me to go over twelve or fifteen miles above Vicksburg." The last sentence reveals clearly enough the madness of attempting to take three of the best ships of the navy to the upper river in falling water. Fortunately the insufficient depth now was above—not below—them, and they were not utterly cut off from the sea. Commander Porter, however, who started down river a week later, in compliance with orders summoning him to Washington, and than whom the navy had no more active nor enterprising officer, wrote back to the flag-officer that if the big ships did not soon return he feared they would have to remain till next year.

Three days after Farragut passed the batteries of Vicksburg, on the 1st of July, the Mississippi flotilla, under the command of Flag-officer Charles H. Davis, joined him from above; having left Memphis only two days before, but favored in their voyage by the current, by competent pilots, and by a draught suited to the difficulties of river navigation. The united squadrons continued together until the 15th of July, lying at anchor near the neck of the promontory opposite Vicksburg; with the exception of the Brooklyn and the two gunboats which had not passed up on the 28th of June. These remained below the works, and on the opposite side of the promontory.

The position of the two flag-officers was about four miles below the mouth of the Yazoo River, a tributary of the Mississippi, which enters the main stream on the east side not far above Vicksburg. It was known to them that there was somewhere in the Yazoo an ironclad ram called the Arkansas; which, more fortunate than the Mississippi at New Orleans, had been hurried away from Memphis just before that city fell into the hands of the United States forces. She was a vessel of between eight hundred and a thousand tons burden, carrying ten guns, which were protected by three inches of railroad iron, backed by bales of compressed cotton firmly braced. Her most dangerous weapon, however, was her ram; but, owing to the lightness and bad construction of the engines, this was not as formidable as it otherwise might have been to the enemy's ships.

So little injury had thus far been done to the United States vessels by the rams of the Confederates that the two flag-officers were probably lulled into a state of over-security, and they allowed their squadrons to lie with too low fires. To this doubtless contributed the more powerful motive of the difficulty to the coal supply incurred by the excessively long line of exposed communications, imposed upon both squadrons by the stubborn persistence of the Navy Department in hurrying the fleets far in advance of any support by the army. Beyond the reach of their guns they could not control the river banks; and, unless they could be present everywhere along the eight hundred miles which separated Memphis from New Orleans, even the narrow strip on either side swept by their cannon was safe at any point only while they were abreast it. The moral effect of their promenade up and down and of their meeting at Vicksburg was accurately weighed by the enemy; and, however it may have imposed upon the Northern people, did nothing to insure the safety of the unarmed vessels upon which supplies depended. This essentially vicious military situation resulted necessarily in a degree of insecurity which could have but one issue—a retreat by both squadrons toward their respective bases, which soon after followed.

Convinced of the inutility of his own presence at Vicksburg, and preoccupied with the risks threatening his squadron from the unguarded state of the river and its dangerous navigation, it is not wonderful that Farragut, who was the senior of the two flag-officers, thought little of the single ironclad vessel in his neighborhood. He was not prone to exaggerate danger, and his experience had not led him to entertain any high opinion of the enemy's rams. To these circumstances he owed one of the most mortifying incidents of his career.

On the 15th of July a reconnoitering expedition was sent into the Yazoo, composed of two vessels of Davis's squadron, accompanied by one of the rams which at that time formed an independent organization upon the upper Mississippi under the command of Colonel Ellet. It was a fortunate move, for to this circumstance was due that the squadrons had any notice of the approach of the Arkansas. The detached vessels met her about six miles within the Yazoo, when a running fight ensued between her and the Carondelet, to the disadvantage of the United States vessel; but the sustained cannonade attracted betimes the attention of the fleet, and the Tyler, a small unarmored boat, after supporting the Carondelet to the best of her ability through the action, preceded the combatants down stream, bringing tidings of the ram's approach. There was not time to raise steam—only to cast loose the guns for action. When the Arkansas reached the fleet her smoke-stack had been so often perforated by the Carondelet's shot that her boilers could scarcely supply any steam. Her speed was thereby reduced to one knot, powerless to ram and scarcely sufficient to steer. At that rate, with the favor also of the current, she passed through the United States vessels, suffering from their successive fires much injury, though not of a vital kind, and took refuge under the guns of Vicksburg. It was a most gallant exploit, fairly comparable in daring to the passage of the Mississippi forts, but resulting in no decisive effect upon the issues of the war.

It became immediately advisable for Farragut to rejoin the three ships which lay below the town, and were consequently in a condition favoring an attack by the ram, whose apparent immunity under the fire of the two squadrons showed her an enemy not to be despised. He determined to follow her down at once, again passing the batteries, and endeavoring to destroy her with the guns of his squadron as it went by. The execution of the plan was set for the late afternoon, and the Mississippi flotilla took up a position to support the movement by engaging the upper batteries. Unfortunately, time was lost in forming the order of battle, and the passage was effected in the dark. The uncertainty of aim thus caused was increased by the precaution of the enemy, who shifted his position after nightfall. Two shots only found her, injuring several of her people and setting fire to the cotton bulwarks. Beyond this she received no injury at this time, but she had been severely shaken by the hammering of the morning. A week later, on the 22d of July, Davis sent down the Essex, one of his heavy ironclads, accompanied by one of Ellet's rams, to attack the Arkansas at her moorings. The effort was unsuccessful, although the enemy's vessel received some further injury. The ram rejoined the upper squadron; but the Essex, from her indifferent speed, was unable to return against the current, exposed unsupported to the fire of all the batteries. She therefore became thenceforth a member of the lower squadron, together with a ram called the Sumter, which had run down with Farragut on the 15th.

On the 20th of the month Farragut had received orders from the Navy Department, dated July 14th, directing him to get the part of his fleet above Vicksburg below that place with as little injury and loss of life as possible. The circumstances that have been narrated caused him to receive this dispatch below the town; and on the 24th, two days after the descent of the Essex, he departed for New Orleans. Davis assured him that the Essex and Sumter should look out for the river between Vicksburg and Baton Rouge. To them were joined three of Farragut's gunboats; and the five vessels took an active part in supporting the garrison of Baton Rouge when an attack was made upon the place by the Confederates on the 5th of August. In this the Arkansas was to have co-operated with the enemy's troops, and she left Vicksburg on the 3d for that purpose; but her machinery broke down, and while lying helpless against the river bank the Essex came in sight. Resistance in her then plight was hopeless. She was set on fire by her commander, the crew escaping to the shore. Farragut himself reached Baton Rouge shortly after this happened. He had with much difficulty succeeded in getting the heavier ships to New Orleans on the 28th of July; and there he had lingered, unwilling to leave the river, though desirous of doing so, until affairs seemed on a reasonably secure basis. The chief element of anxiety was the Arkansas, concerning whose power to harm quite exaggerated notions prevailed. While thus lying before New Orleans word was brought him of the attack on Baton Rouge, and he at once retraced his steps with the Hartford, Brooklyn, and some smaller ships. On the 7th he reached the scene of action, and learned the destruction of the Confederate vessel. The same day he wrote to the Department: "It is one of the happiest moments of my life that I am enabled to inform the Department of the destruction of the ram Arkansas; not because I held the iron-clad in such terror, but because the community did." It must have been an additional element of satisfaction to him that the disappearance from the waters of the Mississippi of the last hostile vessel capable of offensive action released him from the necessity of remaining himself, or of keeping a large force there, during the unhealthy season.

Before leaving Vicksburg the crews of the fleet had suffered severely from the sickness common in that climate. The Brooklyn had sixty-eight sick out of a total of three hundred; and as this proportion was less than in the upper river flotilla, where the sick numbered forty per cent of the total force, it is probable that it fairly represents the general condition of Farragut's ships. Among the troops accompanying the expedition there were but eight hundred fit for duty out of over three thousand. It was not considered well to maintain for a longer time in Baton Rouge the small garrison hitherto stationed there. It had honorably repulsed the enemy's attack; but, in the general cessation of offensive movements by the United States army, the Confederates were continually strengthening their forces on the line of bluffs south of Vicksburg, to the importance of which their attention, never entirely diverted, had been forcibly drawn by the advance of the fleet in the previous months. Fruitless as that ill-judged advance had been, it reminded the enemy of the serious inconvenience they would suffer if the United

States ships could freely patrol that part of the Mississippi, and impressed upon them the necessity of securing a section of it, by which they could have undisturbed communication between the two shores. This could be done by fortifying two points in such strength that to pass them from either direction would involve a risk too great to be lightly undertaken. The points chosen were Vicksburg and Port Hudson, two hundred miles apart, and embracing between them the mouth of the Red River. The latter is the great artery of the region west of the Mississippi, and also, by means of the Atchafalaya Bayou, offers direct communication for light-draught vessels with the Gulf of Mexico. Port Hudson being less than twenty miles from Baton Rouge, the presence in the latter of a small garrison, which could undertake no offensive movement and which there were no troops to re-enforce, became purposeless. On the 16th of August, 1862, the post was abandoned, and the troops occupying it withdrew to New Orleans.

CHAPTER IX.

THE BLOCKADE AND PORT HUDSON.

1862–1863.

OPERATIONS in the Mississippi having now temporarily ceased, Farragut was at liberty to give his undivided attention for a time to the coast blockade. The important harbor of Pensacola had been evacuated by the Confederates in May, less than a month after the capture of New Orleans. Its abandonment was due to want of troops to garrison it properly; the pressure of the United States armies in Kentucky and Tennessee, after the fall of Fort Donelson in the previous February, having necessitated the withdrawal of all men that could be spared from other points. Before the war Pensacola had been the seat of a well-equipped navy yard with a good dry-dock, the only naval station of the United States in the Gulf of Mexico. At the time of the evacuation the buildings in the yard had been destroyed and the dry-dock injured; but the fine harbor, the depth of water—twenty-two feet—that could be carried over the bar, and the nearness of the port to Mobile, the most important center of blockade running, all combined to make it the headquarters of the fleet for repairs and supplies. Farragut arrived there on the 20th of August. Just before leaving

New Orleans he received his commission as rear admiral, dated July 16, 1862. Three other officers were promoted at the same time to the active list of this grade, which had never before existed in the United States; but as Farragut was the senior in rank of the four, he may be said to have been the first officer of the navy to hoist an admiral's flag.

The admiral remained in Pensacola for three months, superintending from there the affairs of his squadron. During this period the harbors of Galveston and of various other smaller ports on the coast of Texas and Lousiana were occupied by detachments of vessels, as the surest way of enforcing the blockade. The admiral had early announced that he should carry on the blockade as far as possible inside; and these successes enabled him to say in December, 1862, that he now held the whole coast except Mobile. During his stay in Pensacola he received a visit from his son, who found him in the best of spirits, all having gone well on the coast; the only mishap having been the success of a Confederate cruiser, the Oreto, in running into Mobile. She had availed herself of her close resemblance to some of the British cruisers in the Gulf to hoist the British flag; and as visits of these vessels to the blockaded ports were authorized and not infrequent, the ruse induced the United States ship that overhauled her to withhold its fire for a few critical moments. During these the Oreto gained so far on the other that, although struck three times by heavy projectiles, she received no vital injury and succeeded in gaining the shelter of the forts.

The period of the admiral's stay in Pensacola was one of the deepest depression to the Union cause, and his letters bear evidence of the anxiety which he shared with all his fellow-countrymen in that time of distress. The reverses of McClellan in the peninsula, followed by the withdrawal of his army from thence and its transference to northern Virginia, the defeats suffered by Pope, and the first invasion of Maryland, occurred either immediately before or during the time that Farragut was in Pensacola. His own bootless expedition up the Mississippi and subsequent enforced retirement conspired also to swell the general gloom; for, although thinking military men could realize from the first that the position into which the fleet was forced was so essentially false that it could not be maintained, the unreflecting multitude saw only the conversion into repulse and disaster of a substantial success, of a conquest as apparently real as it was actually phantasmal. In the West, Grant was so stripped of troops that he feared the possibility of the Union forces being obliged to withdraw behind the Ohio, as they had in the East recrossed the Potomac. "The most anxious period of the war to me," he afterward wrote, "was during the time the army of the Tennessee was guarding the territory acquired by the fall of Corinth and Memphis, and before I was sufficiently re-enforced to take the offensive"—from July 15 to October 15, 1862.

The Confederate forces which confronted Grant in northern Mississippi during these anxious months interposed between him and Vicksburg, and belonged to the department charged with the defenses of the Mississippi river. As they touched Grant, therefore, on the one side, on the other they were in contact with Farragut's command. The summer passed in various movements by them, threatening Grant's position at Corinth, which culminated on the 3d of October in an attack in force. This was repulsed after hard fighting, and re-enforcements to Grant beginning to come in, the Confederates themselves were thrown on the defensive. The approach of winter, bringing with it higher water and healthier weather on the line of the Mississippi, warned them also that the time was at hand when they might have to fight for the control of the water communications, upon which they no longer had, nor could hope to have, a naval force. Reports therefore began to reach the admiral in Pensacola, from the senior naval officer in the river, that the Confederates were with renewed energy building batteries above Baton Rouge and strongly fortifying Port Hudson.

As there seemed no speedy prospect of obtaining the land force, without whose co-operation an attack upon Mobile would be a fruitless enterprise, Farragut felt his proper position was now in the Mississippi itself. Important as was the blockade service, it was of a character safely to be trusted to a subordinate; whereas the strictly military operations of the approaching campaign, whatever shape they might finally take, would be for the control of the river. It therefore behooved the commander-in-chief of the naval forces to be at hand, ready to support in any way that might offer the effort to obtain control of a region of which the water communications were so characteristic a feature. To push far up a narrow and intricate river a force of ships, whose numbers are insufficient even to protect their own communications and insure their coal supplies, is one thing; it is quite another to repair to the same scene of action prepared to support the army by controlling the water, and by establishing in combined action a secure secondary base of operations from which further advances can be made with reasonable certainty of holding the ground gained. There was no inconsistency between Farragut's reluctance of the spring and his forwardness in the autumn. The man who, to secure New Orleans and compass the fall of the forts, had dared to cut adrift from his base and throw his communications to the winds, because he had an object adequate to the risk, was the same who, six weeks later, had testified his anxiety about communications stretched too far and to no purpose; and now, half a year after that reluctant ascent of the river against his better judgment, we find him eagerly planning to go up again, establishing under the protection of the army an advanced base, from which, with the supplies accumulated at it, further movements may be contemplated with a good chance of final success.

On the 14th of November Farragut reported to the Navy Department his return to New Orleans. The Government, however, had taken warning by the fiasco of the previous season; and, far from urging the admiral on, now sought to impress him with the need for caution. As the great object of opening the Mississippi and obtaining control of it remained, and necessarily must remain, the first of the Government's aims in the Southwest, the result of these instructions was to give Farragut the discretion which had before been denied him. He retained fully his convictions of the summer. "I am ready for anything," he writes to the Department, "but desire troops to hold what we get. General Butler urges me to attack Port Hudson first, as he wishes to break up that rendezvous before we go outside. It will take at least five thousand men to take Port Hudson." In the same spirit he writes home, "I am still doing nothing but waiting for the tide of events, and doing all I can to hold what I have"; and again, a week later, "As Micawber says, I am waiting for something to turn up, and in the mean time having patience for the water to rise." Readiness to act, but no precipitation; waiting for circumstances, over which he had no control, to justify acting, may be described as his attitude at this moment.

On the 16th of December the arrival from the north of General Banks to relieve General Butler—an event which took Farragut much by surprise—gave him the opportunity to show at cnce his own ideas of the proper military steps to be taken. Banks had brought re-enforcements with him; and three days after his coming the admiral writes to the Department: "I have recommended to General Banks the occupation of Baton Rouge. . . . It is only twelve or fifteen miles from Port Hudson, and is therefore a fine base of operations. He has approved of the move, and ordered his transports to proceed directly to that point. I ordered Commander James Alden, in the Richmond, with two gunboats, to accompany them and cover the landing." Baton Rouge is on the southernmost of the bluffs which in rapid succession skirt the Mississippi below Vicksburg. With an adequate garrison it became a base of operations from which the army could move against Port Hudson when the time came; and under its protection the colliers and supplies necessary for the naval vessels in the advance could safely remain.

While waiting for the new commander of the army to get fairly settled to work and ready for the combined movement which Farragut was eager to make, the latter was called upon to endure some sharp disappointments. On the 1st of January, 1863, the military forces in Galveston were attacked by Confederate troops, and the naval vessels by a number of river steamboats barricaded with cotton to resist shells fired against them, and loaded with riflemen. The garrison was captured, one of the gunboats blown up by her own officers, and another surrendered after her captain and first lieutenant had been killed on her decks. The other vessels abandoned the harbor. The affair was not only a disaster; it was attended with discreditable circumstances, which excited in the admiral indignation as well as regret. Shortly afterward, two sailing vessels of the squadron, charged with the blockade of Sabine Pass, were also taken by cotton-clad steamers; which to attack availed themselves of a calm day, when the ships were unable to manœuvre. An unsuccessful attempt was made after this to take Sabine Pass; but both that place and Galveston remained in the power of the enemy, and were not regained until the final collapse of the Confederacy. Farragut dispatched one of his most trusted and capable officers, Commodore Henry H. Bell, formerly his chief-of-staff, to re-establish the blockade of Galveston. Arriving off the port toward night, Bell sent one of his detachment, the Hatteras, a light side-wheel iron steamer bought from the merchant service, to overhaul a sail in the offing. Unfortunately, the stranger proved to be the Confederate steamer Alabama, far superior in force to the Hatteras, and after a short engagement the latter was sunk.

All this bad news came in rapid succession, and was closely followed by tidings of the escape from Mobile of the Oreto, which a few months before had eluded the blockading squadron through the daring ruse practiced by her commander. Known now as the Florida, and fitted as a Confederate cruiser, she ran out successfully during the night of January 15th. Here again, though the discredit was less than at Galveston, the annoyance of the admiral was increased by the knowledge that carelessness, or, at the best, bad judgment, had contributed to the enemy's success. From a letter written home at this time by his son, who had not yet returned from the visit begun at Pensacola, it appears that in the intimacy of family life he admitted, and showed by his manner, how keenly he felt the discredit to his command from these events. Though conscious that they were not due to failure on his part to do his utmost with the force given to him, and seeing in the escape of the Oreto a further justification of his own opinion that the lower harbor of Mobile should have been early seized, he nevertheless was "very much worried." This inside view of the effect, visible to those from whom he had no concealments, is supplemented by the description of the admiral's bearing under these reverses given by Captain (now Rear-Admiral) Jenkins, who at this time became his chief-of-staff. "These disasters," he writes, "were sore trials to the admiral, and a less well-poised man would have given way; but they seemed only to give him greater strength of will and purpose. . . . I myself had the misfortune, after months of watching, to see the Oreto run out the first night after I had been relieved of the command of the Oneida and ordered to report to the admiral as his fleet-captain. I had to bear him these bad tidings. Though no stoic, he bore the news as one accustomed to misfortune." It may seem, indeed, that these events, considered individually, were but instances of the hard knocks to be looked for in war, of which every general officer in every campaign must expect to have his share; and this view is undoubtedly true. Nevertheless, occurring in such rapid succession, and all in that part of his extensive command, the blockade, to which at that moment it seemed impossible to give his principal attention, the effect was naturally staggering. His first impulse was to leave the river and repair in person to the scene of disaster in Texas; but reflection soon convinced him that, however unfortunate the occurrences that had taken place there and elsewhere on the coast, they had not the same vital bearing on the issues of the war as the control of the Mississippi, and therefore not an equal claim upon the commander-in-chief.

At the same time, the effect was to intensify the desire to act—to redeem by success the blot which failures had brought upon his command; and the state of affairs elsewhere on the river was becoming such as to justify enterprise by the reasonable hope of substantial results. A series of circumstances which have been often narrated, and nowhere in a more interesting manner than by General Grant in his personal memoirs, had led to the abandonment of the movement by land upon Vicksburg by the Army of the Tennessee, following the Mississippi Central

Railroad. Instead of this original plan of campaign, the Mississippi River was now adopted as the line of advance and of communications. The first move along this new line had been made by General Sherman, who brought with him 32,000 troops, and on the 26th of December, 1862, had landed on the low ground between the mouth of the Yazoo and Vicksburg. On the 29th the army assaulted the works on the hills before them, but were repulsed. Sherman, satisfied that the position there was too strong to be carried, had determined to change his point of attack to the extreme right of the enemy's line, higher up the Yazoo; but the heavy rains which characterized the winter of 1862–'63 in the Mississippi Valley made untenable the ground on which the troops were, and it became necessary to re-embark them. The transports were then moved out into the Mississippi, where they were joined by General McClernand, the senior general officer in the department under Grant himself.

McClernand now decided to attack Arkansas Post, on the Arkansas River, which enters the Mississippi from the west about two hundred miles above Vicksburg. The Post was primarily intended to close the Arkansas and the approach to the capital of the State of the same name; but although fifty miles from the mouth of the river, it was, by the course of the stream, but fifteen by land from the Mississippi. The garrison, being five thousand strong, was thus dangerously placed to threaten the communications by the latter river, upon which the army was to depend during the approaching campaign; and it had already given evidence of the fact by the capture of a valuable transport. This post was reduced on the 11th of January, and McClernand next day started troops up the White River, a tributary of the Arkansas. From this ex-centric movement, which seemed wholly to ignore that Vicksburg and the Mississippi were the objective of the campaign, McClernand was speedily and peremptorily recalled by Grant. The latter, having absolutely no confidence in the capacity of his senior subordinate, could dispossess him of the chief command only by assuming it himself. This he accordingly did, and on the 30th of January joined the army, which was then encamped on the levees along the west bank of the river above Vicksburg.

Serious action on the part of the army, directed by a man of whose vigorous character there could be no doubt, though his conspicuous ability was not yet fully recognized, was evidently at hand; and this circumstance, by itself alone, imparted a very different aspect to any naval enterprises, giving them reasonable prospect of support and of conducing substantially to the great common end. Never in the history of combined movements has there been more hearty co-operation between the army and navy than in the Vicksburg campaign of 1863, under the leadership of Grant and Porter. From the nature of the enemy's positions their forcible reduction was necessarily in the main the task of the land forces; but that the latter were able to exert their full strength, unweakened, and without anxiety as to their long line of communications from Memphis to Vicksburg, was due to the incessant vigilance and activity of the Mississippi flotilla, which grudged neither pains nor hard knocks to support every movement. But, besides the care of our own communications, there was the no less important service of harassing or breaking up those of the enemy. Of these, the most important was with the States west of the Mississippi. Not to speak of cereals and sugar, Texas alone, in the Southwest, produced an abundance of vigorous beef cattle fit for food; and from no other part of the seceded States could the armies on the east banks of the Mississippi be adequately supplied. Bordering, moreover, upon Mexico, and separated from it only by a shoal river into which the United States ships could not penetrate, there poured across that line quantities of munitions of war, which found through the Mexican port of Matamoras a safe entry, everywhere else closed to them by the sea-board blockade. For the transit of these the numerous streams west of the Mississippi, and especially the mighty Red River, offered peculiar facilities. The principal burden of breaking up these lines of supply was thrown upon the navy by the character of the scene of operations—by its numerous water-courses subsidiary to the great river itself, and by the overflow of the land, which, in its deluged condition during the winter, effectually prevented the movement of troops. Herein Farragut saw his opportunity, as well as that of the upper river flotilla. To wrest the control of the Mississippi out of the enemy's hands, by reducing his positions, was the great aim of the campaign; until that could be effected, the patrol of the section between Vicksburg and Port Hudson would materially conduce to the same end.

Over this Farragut pondered long and anxiously. He clearly recognized the advantage of this service, but he also knew the difficulties involved in maintaining his necessary communications, and, above all, his coal. At no time did the enemy cease their annoyance from the river banks. Constant brushes took place between their flying batteries and the different gunboats on patrol duty; a kind of guerrilla warfare, which did not cease even with the fall of Vicksburg and Port Hudson, but naturally attained its greatest animation during the months when their fate was hanging in the balance. The gunboats could repel such attacks, though they were often roughly handled, and several valuable officers lost their lives; but not being able to pursue, the mere frustration of a particular attack did not help to break up a system of very great annoyance. Only a force able to follow—in other words, troops—could suppress the evil. "You will no doubt hear more," the admiral writes on the 1st of February, 1863, "of 'Why don't Farragut's fleet move up the river?' Tell them, Because the army is not ready. Farragut waits upon Banks as to when or where he will go."

Still, even while thus dancing attendance upon a somewhat dilatory general, his plans were maturing; so that when occasion arose he was, as always, ready for immediate action—had no unforeseen decision to make. "The evening of the day (about January 20th) that I reported to him at New Orleans," writes Admiral Jenkins, "he sent everybody out of the cabin, and said: 'I wish to have some confidential talk with you upon a subject which I have had in mind for a long time. . . . I have never hinted it to any one, nor does the department know anything of my thoughts. The first object to be accomplished, which led me to think seriously about it, is to cripple the Southern armies by cutting off their supplies from

Texas. Texas at this time is, and must continue to the end of the war to be, their main dependence for beef cattle, sheep, and Indian corn. If we can get a few vessels above Port Hudson the thing will not be an entire failure, and I am pretty confident it can be done.'" Jenkins naturally suggested that the co-operation of the army by an active advance at the same time would materially assist the attempt. To this, of course, the admiral assented, it being in entire conformity with his own opinion; and several interviews were held, without, however, their leading to any definite promise on the part of General Banks.

Meantime Admiral Porter, who after leaving the mortar flotilla had been appointed to the command of the Mississippi squadron, with the rank of acting rear-admiral, realized as forcibly as Farragut the importance of placing vessels in the waters between Vicksburg and Port Hudson. In the middle of December he was before Vicksburg, and had since then been actively supporting the various undertakings of the land forces. Three days after Grant joined the army, on the 2d of February, the ram Queen of the West ran the Vicksburg batteries from above, and successfully reached the river below. Ten days later, Porter sent on one of his newest ironclads, the Indianola, which made the same passage under cover of night without being even hit, although twenty minutes under fire. The latter vessel took with her two coal barges; and as the experiment had already been successfully tried of casting coal barges loose above the batteries, and trusting to the current to carry them down to the Queen of the West, the question of supplies was looked upon as settled.

The Indianola was very heavily armed, and both the admiral and her commander thought her capable of meeting any force the enemy could send against her.

Unfortunately, on the 14th of February, two days only after the Indianola got down, the Queen of the West was run ashore under a battery and allowed to fall alive into the hands of the enemy. The latter at once repaired the prize, and, when ready, started in pursuit of the Indianola with it and two other steamers; one of which was a ram, the other a cotton-protected boat filled with riflemen. There was also with them a tender, which does not appear to have taken part in the fight. On the night of February 24th the pursuers overtook the Indianola, and a sharp action ensued; but the strength of the current and her own unwieldiness placed the United States vessel at a disadvantage, which her superior armament did not, in the dim light, counterbalance. She was rammed six or seven times, and, being then in a sinking condition, her commander ran her on the bank and surrendered. This put an end to Porter's attempts to secure that part of the river by a detachment. The prospect, that had been fair enough when the Queen of the West was sent down, was much marred by the loss of that vessel; and the subsequent capture of the Indianola transferred so much power into the hands of the Confederates, that control could only be contested by a force which he could not then afford to risk.

The up-river squadron having failed to secure the coveted command of the river, and, besides, transferred to the enemy two vessels which might become very formidable, Farragut felt that the time had come when he not only might but ought to move. He was growing more and more restless, more and more discontented with his own inactivity, when such an important work was waiting to be done. The news of the Queen of the West's capture made him still more uneasy; but when that was followed by the loss of the Indianola, his decision was taken at once. "The time has come," he said to Captain Jenkins; "there can be no more delay. I must go—army or no army." Another appeal, however, was made to Banks, representing the assistance which the squadron would derive in its attempt to pass the batteries from a demonstration made by the army. The permanent works at Port Hudson then mounted nineteen heavy cannon, many of them rifled; but there were reported to be in addition as many as thirty-five field-pieces, which, at the distance the fleet would have to pass, would be very effective. If the army made a serious diversion in the rear, many of these would be withdrawn, especially if Farragut's purpose to run by did not transpire. The advantage to be gained by this naval enterprise was so manifest that the general could scarcely refuse, and he promised to make the required demonstration with eight or ten thousand troops.

On the 12th of March, within a fortnight after hearing of the Indianola affair, Farragut was off Baton Rouge. On the 14th he anchored just above Profit's Island, seven miles below Port Hudson, where were already assembled a number of the mortar schooners, under the protection of the ironclad Essex, formerly of the upper squadron. The admiral brought with him seven vessels, for the most part essentially fighting ships, unfitted for blockade duty by their indifferent speed, but carrying heavy batteries. If the greater part got by, they would present a force calculated to clear the river of every hostile steamer and absolutely prevent any considerable amount of supplies being transferred from one shore to the other.

For the purpose of this passage Farragut adopted a somewhat novel tactical arrangement, which he again used at Mobile, and which presents particular advantages when there are enemies only on one side to be engaged. Three of his vessels were screw steamers of heavy tonnage and battery; three others comparatively light. He directed, therefore, that each of the former should take one of the latter on the side opposite to the enemy, securing her well aft, in order to have as many guns as possible, on the unengaged side, free for use in case of necessity. In this way the smaller vessels were protected without sacrificing the offensive power of the larger. Not only so; in case of injury to the boilers or engines of one, it was hoped that those of her consort might pull her through. To equalize conditions, to the slowest of the big ships was given the most powerful of the smaller ones. A further advantage was obtained in this fight, as at Mobile, from this arrangement of the vessels in pairs, which will be mentioned at the time of its occurrence. The seventh ship at Port Hudson, the Mississippi, was a very large side-wheel steamer. On account of the inconvenience presented by the guards of her wheelhouses, she was chosen as the odd one to whom no consort was assigned

Going up the river toward Port Hudson the course is nearly north; then a bend is reached of over ninety degrees, so that after making the turn the course

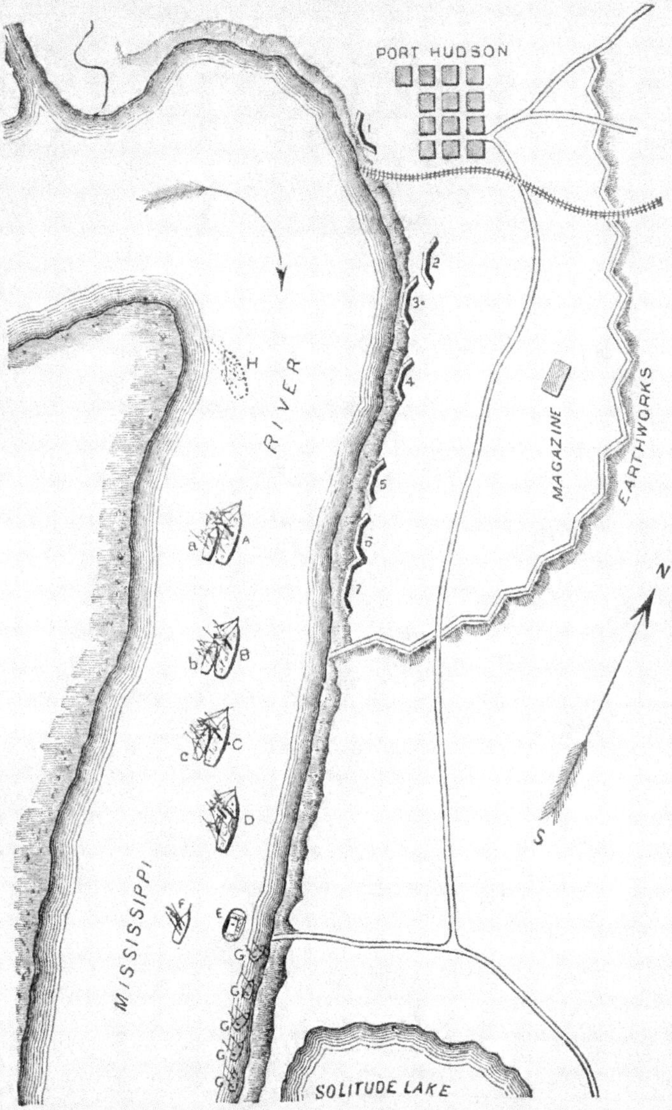

ORDER OF ATTACK ON BATTERIES AT PORT HUDSON, MARCH 14, 1863.

A. Hartford (flagship), Captain James S. Palmer. *a*. Albatross, Lieut.-Com. John E. Hart. B. Richmond, Commander James Alden. *b*. Genesee, Commander W. H. Macomb. C. Monongahela, Captain J. P. McKinstry. *c*. Kineo, Lieut.-Com. John Waters. D. Mississippi, Captain Melancton Smith. E. Essex, Commander C. H. B. Caldwell. F. Sachem, Act. Vol. Lieut. Amos Johnson. G G. Mortar schooners. H. Spot where Mississippi grounded.

for some distance is west-southwest. The town is on the east side, just below the bend. From it the batteries extended a mile and a half down the river, upon bluffs from eighty to a hundred feet high. Between the two reaches, and opposite to the town, is a low, narrow point, from which a very dangerous shoal makes out. The channel runs close to the east bank.

The squadron remained at its anchorage above Profit's Island but a few hours, waiting for the cover of night. Shortly before 10 P. M. it got under way, ranged as follows: Hartford, Richmond, Monongahela, each with her consort lashed alongside, the Mississippi bringing up the rear. Just as they were fairly starting a steamer was seen approaching from down the river, flaring lights and making the loud puffing of the high-pressure engines. The flag-ship slowed down, and the new arrival came alongside with a message from the general that the army was then encamped about five miles in rear of the Port Hudson batteries. Irritated by a delay, which served only to attract the enemy's attention and to assure himself that no diversion was to be expected from the army, the admiral was heard to mutter: "He had as well be in New Orleans or at Baton Rouge for all the good he is doing us." At the same moment the east bank of the river was lit up, and on the opposite point huge bonfires kindled to illumine the scene—a wise precaution, the neglect of which by the enemy had much favored the fleet in the passage of the lower forts.

The ships now moved on steadily, but very slowly, owing to the force of the current. At 11 P. M. the Hartford had already passed the lower batteries, when the enemy threw up rockets and opened fire. This was returned not only by the advancing ships, but also by the ironclad Essex and the mortar schooners, which had been stationed to cover the passage. The night was calm and damp, and the cannonade soon raised a dense smoke which settled heavily upon the water, covering the ships from sight, but embarrassing their movements far more than it disconcerted the aim of their opponents. The flag-ship, being in the advance, drew somewhat ahead of the smoke, although even she had from time to time to stop firing to enable the pilot to see. Her movements were also facilitated by placing the pilot in the mizzen-top, with a speaking tube to communicate with the deck, a precaution to which the admiral largely attributed her safety; but the vessels in the rear found it impossible to see, and groped blindly, feeling their way after their leader. Had the course to be traversed been a straight line, the difficulty would have been much less; but to make so sharp a turn as awaited them at the bend was no easy feat under the prevailing obscurity. As the Hartford attempted it the downward current caught her on the port bow, swung her head round toward the batteries, and nearly threw her on shore, her stem touching for a moment. The combined powers of her own engine and that of the Albatross, her consort, were then brought into play as an oarsman uses the oars to turn his boat, pulling one and backing the other; that of the Albatross was backed, while that of the Hartford went ahead strong. In this way their heads were pointed up stream and they went through clear; but they were the only ones who effected the passage.

The Richmond, which followed next, had reached the bend and was about to turn when a plunging shot upset both safety valves, allowing so much steam to escape that the engines could not be efficiently worked. Thinking that the Genesee, her companion, could not alone pull the two vessels by, the captain of the Richmond turned and carried them both down stream. The Monongahela, third in the line, ran on the shoal opposite to the town with so much violence that the gunboat Kineo, alongside of her, tore loose from the fastenings. The Monongahela remained aground for twenty-five minutes, when the Kineo succeeded in getting her off. She then attempted again to run the batteries, but when near the turn a crank-pin became heated and the engines stopped. Being now unmanageable, she drifted down stream and out of action, having lost six killed and twenty-one wounded. The Mississippi also struck on the shoal, close to the bend, when she was going very fast, and defied every effort to get her off. After working for thirty-five minutes, finding that the other ships had passed off the scene leaving her unsupported, while three batteries had her range and were hulling her constantly, the commanding officer ordered her to be set on fire. The three boats that alone were left capable of floating were used to land the crew on the west bank; the sick and wounded being first taken, the captain and first lieutenant leaving the ship last. She remained aground and in flames until three in the morning, when she floated and drifted down stream, fortunately going clear of the vessels below. At half-past five she blew up. Out of a ship's company of two hundred and ninety-seven, sixty-four were found missing, of whom twenty-five were believed to be killed.

In his dispatch to the Navy Department, written the second day after this affair, the admiral lamented that he had again to report disaster to a part of his command. A disaster indeed it was, but not of the kind which he had lately had to communicate, and to which the word "again" seems to refer; for there was no discredit attending it. The stern resolution with which the Hartford herself was handled, and the steadiness with which she and her companion were wrenched out of the very jaws of destruction, offer a consummate example of professional conduct; while the fate of the Mississippi, deplorable as the loss of so fine a vessel was, gave rise to a display of that coolness and efficiency in the face of imminent danger which illustrate the annals of a navy as nobly as do the most successful deeds of heroism.

Nevertheless, it must be admitted that the failure to pass the batteries, by nearly three fourths of the force which the admiral had thought necessary to take with him, constituted a very serious check to the operations he had projected. From Port Hudson to Vicksburg is over two hundred miles; and while the two ships he still had were sufficient to blockade the mouth of the Red River—the chief line by which supplies reached the enemy—they could not maintain over the entire district the watchfulness necessary wholly to intercept communication between the two shores. Neither could they for the briefest period abandon their station at the river's mouth, without affording an opportunity to the enemy; who was rendered vigilant by urgent necessities which forced him to seize every opening for the passage of stores. From the repulse of five out of the seven ships detailed for the control of the river, it resulted that the enemy's communications, on a line absolutely vital to him, and consequently of supreme strategic importance, were impeded only, not broken off. It becomes, therefore, of interest to inquire whether this failure can be attributed to any oversight or mistake in the arrangements made for forcing the passage—in the tactical dispositions, to use the technical phrase. In this, as in every case, those dispositions should be conformed to the object to be attained and to the obstacles which must be overcome.

The purpose which the admiral had in view was clearly stated in the general order issued to his captains: "The captains will bear in mind that the object is *to run the batteries at the least possible damage to our ships*, and thereby secure an efficient force above, for the purpose of rendering such assistance as may be required of us to the army at Vicksburg, or, if not required there, to our army at Baton Rouge." Such was the object, and the obstacles to its accomplishment were twofold, viz., those arising from the difficulties of the navigation, and those due to the preparations of the enemy. To overcome them, it was necessary to provide a sufficient force, and to dispose that force in the manner best calculated to insure the passage, as well as to entail the least exposure. Exposure is measured by three principal elements—the size and character of the target offered, the length of time under fire, and the power of the enemy's guns; and the last, again, depends not merely upon the number and size of the guns, but also upon the fire with which they are met. In this same general order Farragut enunciated, in terse and vigorous terms, a leading principle in warfare, which there is now a tendency to undervalue, in the struggle to multiply gun-shields and other defensive contrivances. It is with no wish to disparage defensive preparations, nor to ignore that ships must be able to bear as well as to give hard knocks, that this phrase of Farragut's, embodying the experience of war in all ages and the practice of all great captains, is here recalled, "The best protection against the enemy's fire is a well-directed fire from our own guns."

The disposition adopted for the squadron was chiefly a development of this simple principle, combined with an attempt to form the ships in such an order as should offer the least favorable target to the enemy. A double column of ships, if it presents to the enemy a battery formidable enough to subdue his fire, in whole or in part, shows a smaller target than the same number disposed in a single column; because the latter order will be twice as long in passing, with no greater display of gun-power at a particular point. The closer the two columns are together, the less chance there is that a shot flying over the nearer ship will strike one abreast her; therefore, when the two are lashed side by side this risk is least, and at the same time the near ship protects the off one from the projectile that strikes herself. These remarks would apply, in degree, if all the ships of the squadron had had powerful batteries; the limitation being only that enough guns must be in the near or fighting column to support each other, and to prevent several of the enemy's batteries being concentrated on a single ship—a contingency dependent upon the length of the line of hostile guns to be passed. But when, as at Port Hudson, several of the vessels are of feeble gun-power, so that their presence in the fighting column would not re-enforce its fire to an extent at all proportionate to the risk to themselves, the arrangement there adopted is doubly efficacious.

The dispositions to meet and overcome the difficulties imposed by the enemy's guns amounted, therefore, to concentrating upon them the batteries of the heavy ships, supporting each other, and at the same time covering the passage of a second column of gunboats, which was placed in the most favorable position for escaping injury. In principle the plan was the same as at New Orleans—the heavy ships fought while the light were to slip by; but in application, the circumstances at the lower forts would not allow one battery to be masked as at Port Hudson, because there were enemy's works on both sides. For meeting the difficulties of the navigation on this occasion, Farragut seems not to have been pleased with the arrangement adopted. "With the exception of the assistance they might have rendered the ships, if disabled, they were a great disadvantage," he wrote. The exception, however, is weighty; and, taken in connection with his subsequent use of the same order at Mobile, it may be presumed the sentence quoted was written under the momentary recollection of some inconvenience attending this passage. Certainly, with single-screw vessels, as were all his fleet, it was an inestimable advantage, in intricate navigation or in close quarters, to have the help of a second screw working in opposition to the first, to throw the ship round at a critical instant. In the supreme moment of his military life, at Mobile, he had reason to appreciate this advantage, which he there, as here, most intelligently used.

Thus analyzed, there is found no ground for adverse criticism in the tactical dispositions made by Farragut on this memorable occasion. The strong points of his force were utilized and properly combined for mutual support, and for the covering of the weaker elements, which received all the protection possible to give them. Minor matters of detail were well thought out, such as the assignment to the more powerful ship of the weaker gunboat, and the position in which the small vessels were to be secured alongside. The motto that "the best protection against the enemy's fire is a well-directed fire by our own guns" was in itself an epitome of the art of war; and in pursuance of it the fires of the mortar schooners and of the Essex were carefully combined by the admiral with that of the squadron. Commander Caldwell, of the Essex, an exceedingly cool and intelligent officer, reported that "the effect of the mortar fire (two hundred bombs being thrown in one hundred and fifty minutes, from eleven to half-past one) seemed to be to paralyze the efforts of the enemy at the lower batteries; and we observed that their fire was quite feeble compared to that of the upper batteries." Nor had the admiral fallen into the mistake of many general officers, in trusting too lightly to the comprehension of his orders by his subordinates. Appreciating at once the high importance of the object he sought to compass, and the very serious difficulties arising from the enemy's position at Port Hudson and the character of the navigation, he had personally inspected the ships of his command the day before the action, and satisfied himself that the proper arrangements had been made for battle. His general order had already been given to each commanding officer, and he adds: "We conversed freely as to the arrangements, and I found that all my instructions were well understood and, I believe, concurred in by all. After a free interchange of opinions on the subject, every commander arranged his ship in accordance with his own ideas."

In this point the admiral appears to have made a mistake, in not making obligatory one detail which he employed on board the flag-ship. "I had directed a trumpet fixed from the mizzen-top to the wheel on board this ship, as I intended the pilot to take his station in the top, so that he might see over the fog, or smoke, as the case might be. To this idea, and to the coolness and courage of my pilot, Mr. Carrell, I am indebted for the safe passage of this ship past the forts." It may be that the admiral counted upon the vessels being so closed up that the flag-ship would practically serve as the pilot for all. If so, he reckoned without his host, and in this small oversight or error in judgment is possibly to be found a weak point in his preparations; but it is the only one. The failure of the Richmond, his immediate follower, was not in any way due to pilotage, but to the loss of steam by an accidental shot; and it is still a matter of doubt whether the Genesee, her consort, might not have pulled her by. The third in the order, the Monongahela, also failed finally from the heating of a bearing; but as this occurred after being aground for half an hour, with the vigorous working of the engines that naturally ensues under such circumstances, it seems as if her failure must ultimately be traced to the smoke. "The firing had so filled the atmosphere with smoke," wrote her captain, "as to prevent distinguishing objects near by." The loss of the Mississippi was due entirely to an error of the pilot, whatever may have been the cause.

The effect of the appearance above Port Hudson of the Hartford and Albatross is abundantly testified in the correspondence of the day, both Union and Confederate, and justifies beyond dispute this fine conception of Farragut's and the great risk which he took entirely upon his own responsibility. He found, indeed, a ground for his action in an order of the Department dated October 2, 1862,* directing him "to guard the lower part of the Mississippi, especially where it is joined by the Red River," until

---

* The full text of this order was as follows. It committed the department to nothing.

"NAVY DEPARTMENT, *October* 2, 1862.

"SIR: While the Mississippi River continues to be blockaded at Vicksburg, and until you learn from Commander D. D. Porter, who will be in command of the Mississippi squadron, that he has, in conjunction with the army, opened the river, it will be necessary for you to guard the lower part of that river, especially where it is joined by the Red River, the source of many of the supplies of the enemy. I am respectfully, etc.,

"GIDEON WELLES,

"*Secretary of the Navy*."

That five months elapsed between the date of this order and Farragut's action, without anything more definite, shows clearly that the department took no responsibility. On the other hand, it is right to say that it showed a generous appreciation of the effort, and did not complain about the losses.

he heard from Admiral Porter that the latter, in conjunction with the army, had opened the river; but he distrusted the consent of the Secretary to his running the great risk involved in the passage of Port Hudson. As Grant was ordered to take Vicksburg, so was Farragut ordered to blockade the Red River; and as Grant did not notify the commander-in-chief of his final great resolve to cut loose from his base, until it was too late to stop him, so did Farragut keep within his own breast a resolve upon which he feared an interdict. For even after two years of war the department was embarrassed for ships, and the policy of economy, of avoiding risks, the ever fatal policy of a halting warfare, was forced upon it—an impressive illustration of the effect exerted by inadequate preparation upon the operations of war. For lack of ships, Mobile was in 1863 still in the hands of the enemy. "I would have had it long since or been thrashed out of it," wrote Farragut six weeks before Port Hudson. "I feel no fears on the subject; *but they do not wish their ships risked, for fear we might not be able to hold the Mississippi.*" A similar reluctance might be anticipated to expose such valuable vessels as attacked Port Hudson, when their loss was so hard to repair; for only men of the temper of Farragut or Grant—men with a natural genius for war or enlightened by their knowledge of the past—can fully commit themselves to the hazard of a great adventure—can fully realize that a course of timid precaution may entail the greatest of risks.

"Your services at Red River," wrote Admiral Porter to Farragut upon hearing of his arrival above Port Hudson, "will be a godsend; it is worth to us the loss of the "Mississippi," and is at this moment the severest blow that could be struck at the South. They obtain all their supplies and ammunition in that way. . . . The great object is to cut off supplies. For that reason I sent down the Queen of the West and the Indianola. I regret that the loss of the Indianola should have been the cause of your present position." These utterances, which bespeak the relief afforded him at the moment by Farragut's bold achievement, are confirmed by the words written many years later in his History of the Navy. "Farragut in the Hartford, with the Albatross, reached the mouth of the Red River, and Port Hudson was as completely cut off from supplies as if fifty gunboats were there. . . . It was soon seen that the object aimed at had been gained—the works at Port Hudson were cut off from supplies and the fate of the garrison sealed." "I look upon it as of vast importance," wrote General Grant, "that we should hold the river securely between Vicksburg and Port Hudson"; and he undertook to contribute anything that the army could furnish to enable vessels from above to run by Vicksburg, and so supply to Farragut the numbers he needed through the repulse of his own ships.

"The Mississippi is again cut off," wrote to Richmond the Confederate General Pemberton, who commanded the district in which are Vicksburg and Port Hudson, "neither subsistence nor ordnance can come or go"; and the following day, March 20th, the sixth after Farragut's passage, he sends word to General Richard Taylor, on the west shore, "Port Hudson depends almost entirely for supplies upon the other side of the river." "Great God! how unfortunate!" writes, on March

17th, a Confederate commissary in Taylor's department. "Four steamers arrived to-day from Shreveport. One had 300,000 pounds of bacon; three others are reported coming down with loads. Five others are below with full cargoes designed for Port Hudson, but it is reported that the Federal gunboats are blockading the river." As to passing by other points, "it is doubtful whether many cattle ever get through the swamps and bayous through which they are required to pass on this side. As the water declines, I think likely cattle in large quantities can be crossed by swimming, but *at present your prospect of getting supplies from this side is gloomy enough.*" "Early in February," writes Pemberton again, "the enemy succeeded in passing two of his gunboats by our batteries at Vicksburg" (the Indianola and Queen of the West). "This at once rendered the navigation of the Mississippi and Red River dangerous, and from that time it was only by watching opportunities, and at great risk of capture, that supplies could be thrown into Port Hudson and Vicksburg. Nevertheless, large amounts were successfully introduced into both places."

This success, partial as it was, was due, first, to the capture of Porter's detachment, which opened the river again until Farragut came; and, secondly, to the repulse of so large a portion of the latter's squadron. The Hartford and Albatross, though they could close the Red River, could not multiply themselves to cover the great stretch which the admiral had purposed to occupy with seven vessels. Neither was the Albatross of sufficient force to be left by herself at the mouth of the Red River. Farragut therefore moved slowly up the Mississippi, destroying a quantity of stores accumulated upon the levees awaiting transportation, as well as a number of flat-boats; and on the afternoon of the 19th of March he anchored twelve miles below Vicksburg. The following day he moved further up and communicated with General Grant, informing him of the events that had just befallen him and offering any assistance in the power of the two ships. If not needed, he purposed returning to Red River, and asked for coal from either army or navy. Porter was then absent on the Deer Creek expedition, an attempt to get the Mississippi gunboats through the bayou of that name into the Yazoo; whereby, if successful, the Confederate position at Vicksburg would be turned. Grant accordingly undertook to send down coal, which was done by turning adrift in the current of the Mississippi a barge carrying some four hundred tons. This floated by night clear of the enemy's positions, and was picked up by boats from the Hartford.

Farragut had written to Porter of his wish to receive some vessels from above, specifying two rams and an ironclad, with which and his own two vessels he could better carry out his purpose of closing the whole stretch in which he was. He intimated this wish to Grant, who highly approved of it. "I see by Southern papers received yesterday," he wrote to Farragut, "that Vicksburg must depend upon Louisiana, or west of the Mississippi, for supplies. Holding Red River from them is a great step in the direction of preventing this, but it will not entirely accomplish the object. New Carthage (twenty miles below Vicksburg, on the west bank) should be held, and it seems to me that in addition we should have sufficient vessels below to patrol the whole river from Warrenton (ten miles below Vicksburg) to the Red River. I will have a consultation with Admiral Porter on this subject. I am happy to say the admiral and myself have never yet disagreed upon any policy." In the absence of Porter, General Ellet determined to send down two of the Ellet rams, which made their dash on the morning of March 25, displaying all the daring, but unfortunately also much of the recklessness, which characterized that remarkable family. Starting near dawn, on a singularly clear night, they were surprised by daylight still under fire. One, being very rotten, was shattered to pieces by a shell exploding her boilers. The other was disabled, also by a shell in the boilers, but, being stronger, drifted down with the current and reached Farragut safely. She was soon repaired, and was an addition to his force.

While lying below Vicksburg the admiral transferred to Porter's care, for passage north by the Mississippi River, his son and only child, who had been with him since the summer stay in Pensacola. They had passed the batteries at Port Hudson together, the bearing of the boy in that hot contest approving itself to the father, who, despite his anxiety, could not bring himself to accept the surgeon's suggestion to send him below, out of harm's way. "I am trying to make up my mind to part with Loyall," he wrote to his wife, "and to let him go home by way of Cairo. I am too devoted a father to have my son with me in troubles of this kind. The anxieties of a father should not be added to those of the commander."

On the 27th of March the Hartford started again down river, accompanied by the Albatross and the

Ellet ram Switzerland. On the 2d of April the little squadron anchored off the mouth of the Red River, having on its passage down again destroyed a number of skiffs and flat-boats used for transporting stores. Warned by the fate of the Indianola, the admiral left nothing undone to ensure the absolute safety of the flag-ship; for, though her powerful armament and numerous crew gave her a great superiority over any number of river vessels, granting her room to manœuvre, the difficulties of the river and the greatness of the stake to both parties made it imperative to take no needless risks. As a protection against rams, large cypress logs were hung around the ship about a foot above the water line, where they would both resist penetration and also give time for the elasticity of the frame of a wooden vessel to take up the blow. Against boarding, elaborate preparations were made, which would prevent a steamer attempting it from getting nearer than twenty feet to the side, where she would remain an easy victim to the shell and grape of the Hartford's guns.

From the 2d to the 30th of April Farragut remained in the neighborhood of the Red River, between its mouth and Port Hudson. Cut off by the batteries of the place, and by the prevalence of guerrillas on the west bank, from all usual means of communication with General Banks and his own squadron, he contrived to get a letter down by the daring of his secretary, Mr. Edward C. Gabaudan; who was set adrift one night in a skiff ingeniously covered with drift brush, and, thus concealed, floated undiscovered past the enemy's guards. The small number of his vessels prevented his extending his blockade as far as he wished; but in closing the Red River he deprived the enemy of by far the best line they possessed, and he destroyed a quantity of stores and boats.

In the mean time diverse and important events were concurring to release him from his position of isolation. Toward the end of March General Grant, who had for some time abandoned all expectation of turning Vicksburg by its right flank, began the celebrated movement down the west side of the Mississippi; whence he crossed to the east bank at Bruinsburg, and fought the campaign which ended by shutting up Pemberton and his army within the lines of the place. In furtherance of this plan, Porter himself, with a large body of his ships, ran the batteries at Vicksburg on the night of April 16. The fleet then kept pace with the necessarily slow progress of the army, encumbered with trains, through the roads heavy with the mire of the recent overflow. On the 29th of April the Mississippi squadron fought a sharp engagement with the Confederate batteries at Grand Gulf, which they could not reduce; and the following day Grant's army crossed the river.

While these events were bringing the Mississippi squadron into that part of the river which Farragut had aimed to control, other movements were leading to his assistance some of the lighter vessels of his own command. After the naval action at Port Hudson, Banks had temporarily abandoned his designs upon that post in favor of operations west of the Mississippi by the Bayous Teche and Atchafalaya, the latter of which communicates with the Red River a few miles above its mouth. This movement was accompanied by a force of four gunboats, under the command of Lieutenant-Commander A. P. Cooke, of the Estrella, which captured a post on the Atchafalaya called Butte à la Rose, on the 20th of April, the same day that Opelousas, sixty miles from Alexandria, was entered by the army. The latter pressed on toward Alexandria, while the gunboats pushed their way up the Atchafalaya. On the first of May two of them, the Estrella and Arizona, passed into the Red River, and soon afterward joined the Hartford.

Three days later Admiral Porter arrived with several of his fleet and communicated with Farragut. The next day, May 5th, Porter went up the Red River and pushed rapidly toward Alexandria, which was evacuated, its stores being removed to Shreveport, three hundred and fifty miles farther up.

Farragut now felt that his personal presence above Port Hudson was no longer necessary. The Mississippi was ultimately to become the command of Porter, whose vessels were especially fitted for its waters; and that admiral was now at liberty to give his full attention below Vicksburg. On the other hand, his own squadron in the lower river and on the blockade demanded a closer attention than he could give from his isolated station. Accordingly, on the 6th of May he transferred the command to Commodore Palmer, of the Hartford, with whom he left the Albatross, Estrella, and Arizona to intercept communications between the two banks of the Mississippi below Red River; while he himself returned by one of the bayous to New Orleans, reaching there on the 11th.

Thus ended Farragut's brilliant strategic movement against the communications of Vicksburg and Port Hudson, and through them against the intercourse of the Confederacy with its great Western storehouse, over which the two fortresses stood guard. It was a movement which, though crippled from the beginning by a serious disaster on the battle-field, was conceived in accordance with the soundest principles of the art of war. Its significance has been obscured and lost in the great enterprise initiated a month later by General Grant, and solidly supported by the navy under Porter; whose co-operation, Grant avows, was absolutely essential to the success—nay, even to the contemplation of such an undertaking.* In this combined movement, identical in principle with that of Farragut, Porter, in executing his part, had the current with instead of against him. Had circumstances delayed or prevented Grant's advance by the west bank of the Mississippi—had he, for instance, been enabled by one of the abortive bayou expeditions to penetrate north of Vicksburg—Farragut's action would have been no more sound nor bold, but its merits would have been far more perceptible to the common eye. Re-enforcements must have been sent him; and around his flagship would have centered a force that would have choked the life out of Vicksburg and Port Hudson.

Because rightly aimed, this daring campaign was not frustrated even by the disasters of the night action. It is distinguished from the unhappy fiasco of the year before by all the difference between a fitting and an unfitting time—by all that separates a clear appreciation of facts from a confused impression of possibilities. In 1862 Farragut was driven up the river against his own judgment, seeing no prospect

* *Personal Memoirs of U. S. Grant*, vol. i, p. 461.

of tangible or permanent results. In 1863 he went on his own responsibility, because he saw that in the then condition of affairs, with the armies gathering at both ends of the line, the movement he made would not only be successful in itself, but would materially conduce to the attainment of the common end. It is significant of his true military insight that neither depreciation nor disaster shook his clear convictions of the importance of his work. "Whether my getting by Port Hudson was of consequence or not," he wrote chaffingly in reference to some slighting comments in a Southern newspaper, "if Pollard's stomach were as tightly pinched for food as theirs at Port Hudson and Vicksburg have been since I shut up Red River, he would know how to value a good dinner and a little peace." In soberer style he wrote to his home: "We have done our part of the work assigned to us, and all has worked well. *My last dash past Port Hudson was the best thing I ever did*, except taking New Orleans. It assisted materially in the fall of Vicksburg and Port Hudson."

Farragut remained but a short time in New Orleans, and was soon again at the front; joining the vessels of his squadron lying near, but below, Port Hudson. After entering Alexandria on the 7th of May, General Banks moved down with his army to the Mississippi, which he crossed five or six miles above Port Hudson. General Augur, of his command, at the same time moved up from Baton Rouge, the two divisions meeting on the 23d of May, and immediately investing Port Hudson. An assault was made on the 27th; but proving unsuccessful, regular siege operations were begun. The mortar schooners and the Essex supported them by constant bombardment, and the navy furnished and manned a battery of four nine-inch Dahlgren guns.

While contributing thus conspicuously to the immediate furtherance of the siege, the most essential work of the navy, here as in the upper Mississippi, was in the maintenance of the communications, which were wholly by the river, as well as in assuring the safety of New Orleans, then stripped of all the troops that could be spared. The danger of two points like Vicksburg and Port Hudson, both of such vital importance, and both being besieged at the same time, aroused every latent energy of the Confederacy, and set in motion every armed man of whom it could dispose. To divert and distract the attention of the Union generals, to induce them to abandon their efforts or diminish the forces at the front, no means were so ready nor so sure as an attack upon their communications, or a threat directed against their base. To make these insecure, is like mining the foundations of a building. Here the navy removed every substantial cause of anxiety by its firm support, and by the rapidity with which its heavy guns were brought to sustain every point attacked. Under such diligent guardianship the barrier of the Mississippi remained impassable; and although a transport might now and again be arrested and forced to surrender, such an occasional annoyance could not by the most uneasy general be magnified into a serious menace to his communications. The active Confederate general, Richard Taylor, in command of the district west of the river, stripped all his posts to concentrate an effort along the right bank, which, by disturbing Banks, might make a favorable diversion for Port Hudson; and loud talk was made of an attack upon New Orleans itself, favored by a rising among the citizens, still heartily attached to the Southern cause. The powerful vessels kept before the city by Farragut effectually disposed of any chance of such an attempt, although much anxiety was felt by General Emory, in command of the station, and confident expectation was plainly discernible on the faces of the towns-people. The Confederates, however, did for a season control the west side of the river, appearing before Donaldsonville and Plaquemine, where were posts of United States troops. These were saved by the prompt appearance of gunboats, which followed the movements of the enemy; but the report of them brought Farragut down in person, and elicited from him a remonstrance to Banks for leaving upon the west bank, inadequately sustained, heavy guns which, if they fell into the hands of the Confederates, might convert a menace into a serious embarrassment. A few days later, at midnight of June 27th, the enemy attacked Donaldsonville in force. The storming party succeeded in entering the works, but the three gunboats which Farragut had stationed there opened so heavy a fire upon the supports that these broke and fled; and those in advance, being unsustained, were made prisoners.

A few days later Farragut summoned his chief-of-staff, Captain Thornton A. Jenkins, to relieve him at Port Hudson, as he felt his own presence necessary at New Orleans. Jenkins started up in the Monongahela, a heavy corvette commanded by Captain Abner Read, having in company two small transports with needed supplies. The enemy, despite the repulse at Donaldsonville, remained in the neighborhood, and had established a battery of field-guns a few miles below at a bend in the river. By these the Monongahela was attacked and pretty severely handled for a few moments. Her captain, an officer of distinguished courage and enterprise, was mortally wounded, and Captain Jenkins slightly so. These two affairs sufficiently indicate the character of the enemy's operations on the west bank of the Mississippi at this time. They did not in the least succeed in shaking the grip of the Union army before Port Hudson, nor did they entirely cease with the surrender of the place. That they did so little harm, with the enemy in nearly undisputed command in the regions west of the river, was due to the navy, whose mobility exceeded that of their troops.

Vicksburg surrendered on the 4th of July, 1863, and its fall was followed by that of Port Hudson on the 9th of the same month. Farragut then wrote to Porter, and turned over to him the command in all the Mississippi Valley above New Orleans. On the 1st of August Porter himself arrived off the city in his flag-ship, and the two admirals had an interview on the scene of their former exploits. The same afternoon Farragut sailed in the Hartford for the North, to enjoy a brief respite from his labors during the enervating autumn months of the Gulf climate. Though now sixty-two years old, he retained an extraordinary amount of vitality, and of energy both physical and moral; but nevertheless at his age the anxieties and exposure he had to undergo tell, and had drawn from him, soon after his return from above Port Hudson, the expressive words, "I am growing old fast, and need rest." On the 10th of August the flag-ship anchored in New York, after a passage of nine days.

The admiral remained in the North until the first of the following year. His own ship, and her powerful sisters, the Richmond and Brooklyn, were in need of extensive repairs before they could be considered again fit for winter service in the Gulf. The Hartford was in better condition than the other two, being uninjured below the water line, but the severe actions through which she had passed were proved by the scars, two hundred and forty in number, where she had been struck by shot or shell.

CHAPTER X.

MOBILE.

1864.

By the fall of the last and most powerful of the Confederate strongholds upon the Mississippi, and the consequent assertion of control by the United States Government over the whole of the great water course, was accomplished the first and chief of the two objects toward which Farragut was to cooperate. After manifold efforts and failures, the combined forces of the United States had at last sundered the Confederacy in twain along the principal one of those natural strategic lines which intersected it, and which make the strength or the weakness of States according as they are able or unable to hold them against an enemy. Of the two fragments, the smaller was militarily important only as a feeder to the other. Severed from the body to which they belonged, the seceded States west of the Mississippi sank into insignificance; the fire that had raged there would smoulder and die of itself, now that a broad belt which could not be passed interposed between it and the greater conflagration in the East.

It next became the task of the Union forces to hold firmly, by adequate defensive measures, the line they had gained; while the great mass of troops heretofore employed along the Mississippi in offensive operations were transferred farther east, to drive yet another column through a second natural line of cleavage from Nashville, through Georgia, to the Gulf or to the Atlantic seaboard. How this new work was performed under the successive leadership of Rosecrans, Grant, and Sherman, does not fall within the scope of the present work. Although the light steamers of the Mississippi squadron did good and often important service in this distant inland region, the river work of Farragut's heavy sea-going ships was now over. In furtherance of the great object of opening the Mississippi, they had left their native element, and, braving alike a treacherous navigation and hostile batteries, had penetrated deep into the vitals of the Confederacy. This great achievement wrought, they turned their prows again seaward. The formal transfer to Admiral Porter of the command over the whole Mississippi and its tributaries, above New Orleans, signalized the fact that Farragut's sphere of action was to be thenceforth on the coast; for New Orleans, though over a hundred miles from the mouth of a tideless river, whose waters flow ever downward to the sea, was nevertheless substantially a sea-coast city.

As the opening of the Mississippi was the more important of the two objects embraced in Farragut's orders, so did it also offer him the ampler field for the display of those highest qualities of a general officer which he abundantly possessed. The faculty of seizing upon the really decisive points of a situation, of correctly appreciating the conditions of the problem before him, of discerning whether the proper moment for action was yet distant or had already arrived, and of moving with celerity and adequate dispositions when the time did come—all these distinctive gifts of the natural commander-in-chief had been called into play, by the difficult questions arising in connection with the stupendous work of breaking the shackles by which the Confederates held the Mississippi chained. The task that still remained before him, the closing of the Confederate seaports within the limits of his command, though arduous and wearisome, did not make the same demand upon these more intellectual qualities. The sphere was more contracted, more isolated. It had fewer relations to the great military operations going on elsewhere, and, being in itself less complex, afforded less interest to the strategist. It involved, therefore, less of the work of the military leader which was so congenial to his aptitudes, and more of that of the administrator, to him naturally distasteful.

Nevertheless, as the complete fulfilment of his orders necessitated the reduction of a fortified seaport, he found in this undertaking the opportunity for showing a degree of resolution and presence of mind which was certainly not exceeded—perhaps not even equaled—in his previous career. At Mobile it was the tactician, the man of instant perception and ready action, rather than he of clear insight and careful planning, that is most conspicuous. On the same occasion, with actual disaster incurred and imminent confusion threatening his fleet, combined with a resistance sturdier than any he had yet encountered, the admiral's firmness and tenacity rose equal to the highest demand ever made upon them. In the lofty courage and stern determination which plucked victory out of the very jaws of defeat, the battle of Mobile Bay was to the career of Farragut what the battle of Copenhagen was to that of Nelson. Perhaps we may even say, borrowing the words of an eloquent French writer upon the latter event, the battle of Mobile will always be in the eyes of seamen Farragut's surest claim to glory.*

Up to the time of Farragut's departure for the North, in August, 1863, the blockade of the Gulf sea-coast within the limits of his command, though technically effective, had for the most part only been enforced by the usual method of cruising or anchoring off the entrances of the ports. Such a watch, however, is a very imperfect substitute for the iron yoke that is imposed by holding all the principal harbors, the gateways for communication with the outer world. This was clearly enough realized; and the purpose of Farragut, as of his Government, had been so to occupy the ports within his district. At one time, in December, 1862, he was able to say exultingly that he did so hold the whole coast, except Mobile; but the disasters at Galveston and Sabine Pass quickly intervened, and those ports remained thenceforth in the hands of the enemy. On the Texas coast, however, blockade-running properly so called—the entrance, that is, of blockaded Confederate harbors—was a small matter compared with the flourishing contraband trade carried on through the Mexican port Matamoras and across the Rio Grande. When Farragut's lieutenant, Commodore Henry H.

* "The campaign of the Baltic will always be in the eyes of seamen Nelson's fairest claim to glory. He alone was capable of displaying such boldness and such perseverance; he alone could face the immense difficulties of that enterprise and triumph over them."—Jurien de la Gravière, *Guerres Maritimes.*

Bell, visited this remote and ordinarily deserted spot in May, 1863, he counted sixty-eight sails at anchor in the offing and a forest of smaller craft inside the river, some of which were occupied in loading and unloading the outside shipping; to such proportions had grown the trade of a town which neither possessed a harbor nor a back country capable of sustaining such a traffic. Under proper precautions by the parties engaged, this, though clearly hostile, was difficult to touch; but it also became of comparatively little importance when the Mississippi fell.

Not so with Mobile. As port after port was taken, as the lines of the general blockade drew closer and closer, the needs of the Confederacy for the approaching death-struggle grew more and more crying, and the practicable harbors still in their hands became proportionately valuable and the scenes of increasing activity. After the fall of New Orleans and the evacuation of Pensacola, in the spring of 1862, Mobile was by far the best port on the Gulf coast left to the Confederates. Though admitting a less draught of water than the neighboring harbor of Pensacola, it enjoyed the advantage over it of excellent water communications with the interior; two large rivers with extensive tributary systems emptying into its bay. Thanks to this circumstance, it had become a place of very considerable trade, ranking next to New Orleans in the Gulf; and its growing commerce, in turn, reacted upon the communications by promoting the development of its railroad system. The region of which Mobile was the natural port did not depend for its importance only upon agricultural products; under somewhat favorable conditions it had developed some manufacturing interests in which the Southern States were generally very deficient, and which afterward found active employment in the construction of the Tennessee, the most formidable ironclad vessel built by the Confederates.

For all these reasons the tenure of Mobile became a matter of serious consequence to the enemy; and, as Farragut had from the first foreseen, they made active use of the respite afforded them by the unfortunate obstinacy of the Navy Department in refusing him permission to attack after New Orleans fell. The enterprise then was by no means as difficult as the passage of the Mississippi forts just effected; and once captured, the holding of the harbor would require only the small number of troops necessary to garrison the powerful masonry fort which commanded the main ship channel, supported by a naval force much less numerous than that required to blockade outside. The undertaking was therefore not open to the objection of unduly exposing the troops and ships placed in unfortified or poorly fortified harbors, which received such a sad illustration at Galveston; but it was dropped, owing, first, to the preoccupation of the Government with its expectations of immediately reducing the Mississippi, and afterward to the fear of losing ships which at that time could not be replaced. Hesitation to risk their ships and to take decisive action when seasonable opportunity offers, is the penalty paid by nations which practise undue economy in their preparations for war. When at last it became urgent to capture Mobile before the powerful ironclad then building was completed, the preparations of the defense were so far advanced that ironclad vessels were needed for the attack; and before these could be, or at least before they were, supplied, the Tennessee, which by rapid action might have been forestalled like the similar vessel at New Orleans, was ready for battle. Had she been used with greater wisdom by those who directed her movements, she might have added very seriously to the embarrassment of the United States admiral.

When Farragut, after an absence of nearly six months, returned to his station in January, 1864, it was with the expectation of a speedy attack upon Mobile. On his way to New Orleans he stopped off the bar, and on the 20th of January made a reconnaissance with a couple of gunboats, approaching to a little more than three miles from the forts commanding the entrance. He then reported to the department that he was satisfied that, if he had one ironclad, he could destroy the whole of the enemy's force in the bay, and then reduce the forts at leisure with the co-operation of about five thousand troops. "But without ironclads," he added, "we should not be able to fight the enemy's vessels of that class with much prospect of success, as the latter would lie on the flats, where our ships could not get at them. By reference to the chart you will see how small a space there is for the ships to manœuvre. Wooden vessels can do nothing with the ironclads, unless by getting within one or two hundred yards, so as to ram them or pour in a broadside." He repeats the information given by a refugee, that the ironclad Nashville would not be ready before March, and that the Confederate admiral announced that when she was he would raise the blockade. "It is depressing," he adds, "to see how easily false reports circulate, and in what a state of alarm the community is kept by the most absurd rumors. If the Department could get one or two ironclads here, it would put an end to this state of things and restore confidence to the people of the ports now in our possession. I feel no apprehension about Buchanan's raising the blockade; but, with such a force as he has in the bay, it would be unwise to take in our wooden vessels without the means of fighting the enemy on an equal footing." Having made this reconnaissance, he went on to New Orleans, arriving there January 22d.

It appears, therefore, that, regarded as a naval question, Farragut considered the time had gone by for an attempt to run the forts of Mobile Bay, and that it would not return until some ironclads were furnished him by the Department. The capture of the forts he at no time expected, except by the same means as he had looked to for the reduction of those in the Mississippi—that is, by a combined military and naval operation. In both cases the navy was to plant itself across the enemy's communications, which it could do by running the gantlet of his guns. It then remained for the land forces either to complete the investment and await their fall by the slow process of famine, or to proceed with a regular siege covered by the fleet. Without the protection of the ships in the bay, the army would be continually harassed by the light gunboats of the enemy, and very possibly exposed to attack by superior force. Without the troops, the presence of the ships inside would be powerless to compel the surrender of the works, or to prevent their receiving some supplies. But in the two years that had very nearly elapsed since Farragut, if permitted his own wish, would have attacked, the strengthening of the works and the introduction of the ironclads had materially altered the question. He was, it is true, misinformed as to the readiness of the latter. The vessels that were dignified by that name when he first returned to his station, took no part in the defense, either of the bay or, later, of the city. He was deceived, probably, from the fact that the Confederates themselves were deceived, with the exception of a few who had more intimate knowledge of their real value; and consequently the reports that were brought off agreed in giving them a character which they did not deserve.

An attack upon Mobile had been a cherished project with General Grant after the fall of Vicksburg. It was to that—and not to the unfortunate Red River expedition of 1864—that he would have devoted Banks's army in the Southwest; moving it, of course, in concert with, so as to support and be supported by, the other great operations which took place that year—Sherman's advance upon Atlanta and his own against Richmond. It was to Mobile, and not to Savannah, that he first looked as the point toward which Sherman would act after the capture of Atlanta; the line from Atlanta to Mobile would be that along which, by the control of the intervening railroad systems, the Confederacy would again be cleft in twain, as by the subjugation of the Mississippi. For this reason chiefly he had, while still only commander of the Army of the Tennessee, and before he succeeded to the lieutenant-generalship and the command of all the armies, strenuously opposed the Red River expedition; which he looked upon as an ex-centric movement, tending rather to keep alive the war across the Mississippi, which would fade if left alone, and likely to result in the troops engaged not getting back in time or in condition to act against Mobile.

As Grant feared, so it happened. The expedition being already organized and on the point of starting when he became commander-in-chief, he allowed it to proceed; but it ended in disaster, and was the cause of forty thousand good troops being unavailable for the decisive operations which began two months later. Not until the end of July could a force be spared even for the minor task of reducing the Mobile forts; and until then Farragut had to wait in order to attack to any purpose. By the time the army in the Southwest, in the command of which General Canby relieved Banks on the 20th of May, was again ready to move, Sherman had taken Atlanta, Hood had fallen upon his communications with Chattanooga, and the famous march to the sea had been determined. Farragut's battle in Mobile Bay therefore did not prove to be, as Grant had hoped, and as his passage of the Mississippi forts had been, a step in a series of grand military operations, by which the United States forces should gain control of a line vital to the Confederacy, and again divide it into two fragments. It remained an isolated achievement, though one of great importance, converting Mobile from a maritime to an inland city, putting a stop to all serious blockade-running in the Gulf, and crushing finally the enemy's ill-founded hopes of an offensive movement by ironclads there equipped.

The city of Mobile is itself some thirty miles from the Gulf, near the head of a broad but gener-

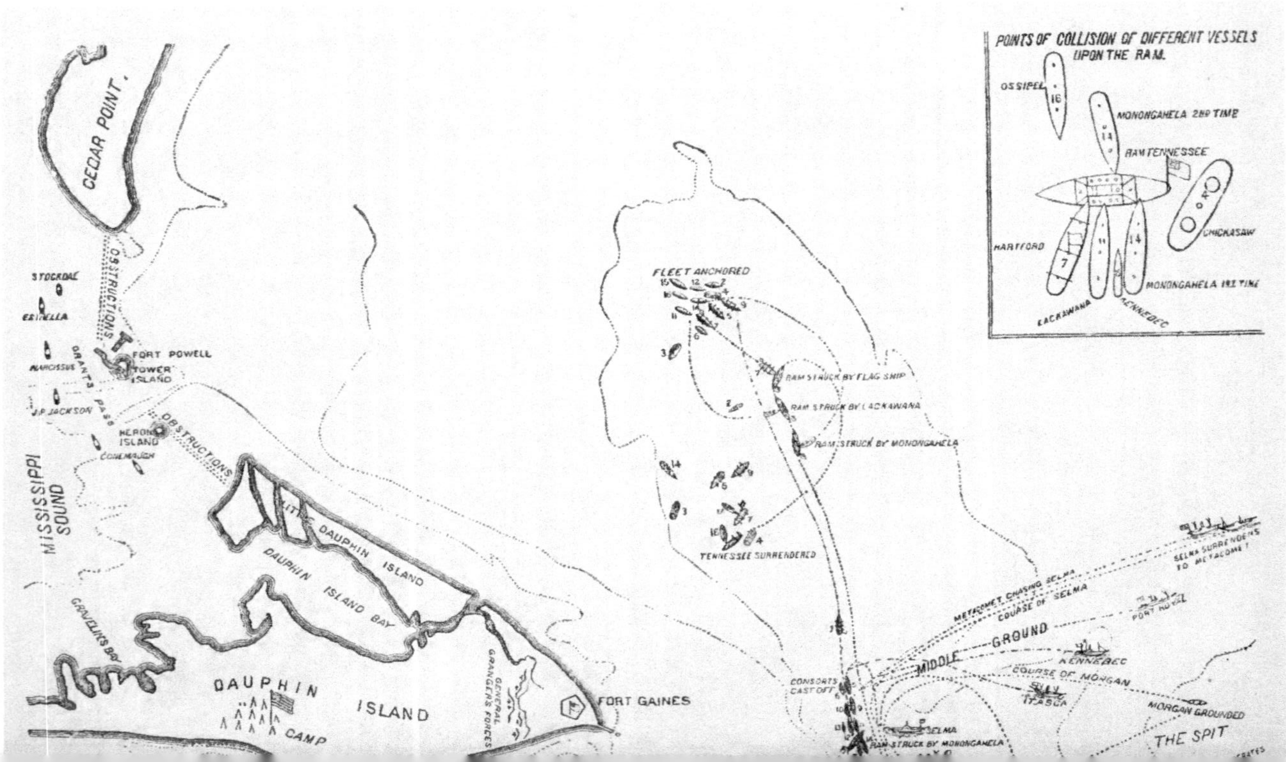
POINTS OF COLLISION OF DIFFERENT VESSELS UPON THE RAM.
OSSIPEE
MONONGAHELA 2ND TIME
RAM TENNESSEE
CHICKASAW
HARTFORD
MONONGAHELA 1ST TIME
LACKAWANA
KENNEBEC
CEDAR POINT.
OBSTRUCTIONS
STOCKDALE
ESTRELLA
NARCISSUS
FORT POWELL
TOWER ISLAND
GRANTS PASS
J.P. JACKSON
HERON ISLAND
CONEMAUGH
OBSTRUCTIONS
MISSISSIPPI SOUND
LITTLE DAUPHIN ISLAND
DAUPHIN ISLAND BAY
GRAVELINE BAY
DAUPHIN ISLAND
CAMP
GENERAL GRANGER'S FORCES
FORT GAINES
FLEET ANCHORED
RAM STRUCK BY FLAG SHIP
RAM STRUCK BY LACKAWANA
RAM STRUCK BY MONONGAHELA
TENNESSEE SURRENDERED
CONSORTS CAST OFF
SELMA
RAM STRUCK BY MONONGAHELA
METACOMET CHASING SELMA
COURSE OF SELMA
MIDDLE GROUND
KENNEBEC
COURSE OF MORGAN
ITASCA
MORGAN GROUNDED
THE SPIT
PORT ROYAL
SELMA SURRENDERS TO METACOMET

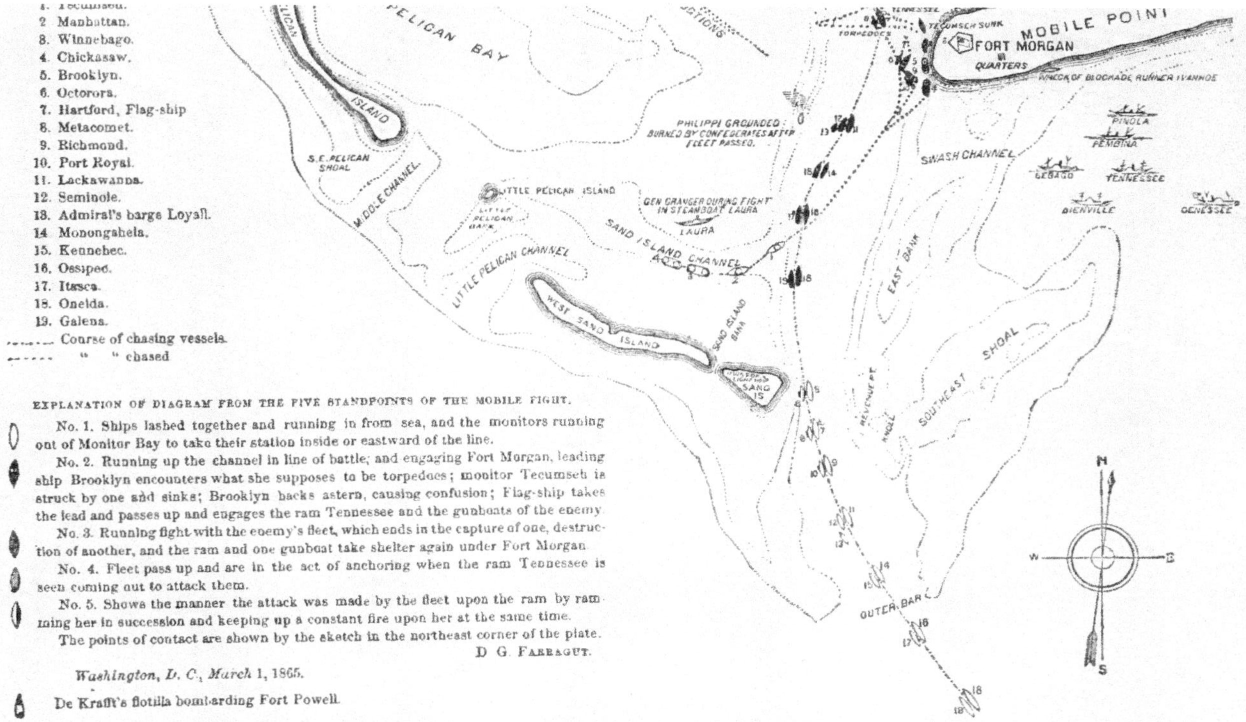

Entrance of Rear-Admiral Farragut's Fleet into Mobile Bay, August 5, 1864.

ally shallow bay which bears the same name. The principal entrance from the Gulf is between Mobile Point—a long, narrow, sandy beach which projects from the east side of the bay—and Dauphin Island, one of a chain which runs parallel to the coast of Mississippi and encloses Mississippi Sound. At the end of Mobile Point stands Fort Morgan, the principal defense of the bay, for the main ship channel passes close under its guns. At the eastern end of Dauphin Island stood a much smaller work, called Fort Gaines. Between this and Fort Morgan the distance is nearly three miles; but a bank of hard sand making out from the island prevents vessels of any considerable size approaching it nearer than two miles. Between Dauphin Island and the mainland there are some shoal channels, by which vessels of very light draft can pass from Mississippi Sound into the bay. These were not practicable for the fighting vessels of Farragut's fleet; but a small earthwork known as Fort Powell had been thrown up to command the deepest of them, called Grant's Pass.

The sand bank off Dauphin Island extends south as well as east, reaching between four and five miles from the entrance. A similar shoal stretches out to the southward from Mobile Point. Between the two lies the main ship channel, varying in width from seven hundred and fifty yards, three miles outside, to two thousand, or about a sea mile, abreast Fort Morgan. Nearly twenty-one feet can be carried over the bar; and after passing Fort Morgan the channel spreads, forming a hole or pocket of irregular contour, about four miles deep by two wide, in which the depth is from twenty to twenty-four feet. Beyond this hole, on either side the bay and toward the city, the water shoals gradually but considerably, and the heavier of Farragut's ships could not act outside of its limits. The Confederate ironclad Tennessee, on the contrary, drawing but fourteen feet, had a more extensive field of operations open to her, and, from the gradual diminution of the soundings, was able to take her position at a distance where the most formidable of her opponents could neither follow her nor penetrate her sides with their shot.

Between the city and the lower bay there were extensive flats, over which not even the fourteen feet of the Tennessee could be taken; and these in one part, called Dog River Bar, shoaled to as little as nine feet. To bring the Tennessee into action for the defense of the entrance and of the lower bay, it was necessary to carry her across these flats —an undertaking requiring both time and mechanical appliances, neither of which would be available if an enemy were inside to molest the operations. As the Tennessee was distinctly the most formidable element in the dangers Farragut had to encounter, and as the character of the soundings gave her a field of action peculiarly suited to utilize her especial powers, which consisted in the strength of her sides and the long range of her heavy rifled guns, it was particularly desirable to anticipate her crossing the upper bar by the fleet itself crossing the lower. That done, the Tennessee was reduced to impotence. It was not done, for two reasons. First, the Navy Department did not send the ironclads which Farragut demanded; and second, the army in the Southwest, having wasted its strength in a divergent operation, was unable to supply the force necessary to reduce Fort Morgan. That the delay was not productive of more serious consequences was due to the impatience or recklessness of the Confederate admiral, and to the energy with which Farragut seized the opportunity afforded by his mistake.

Six months passed before the moment for decisive action arrived. Though devoid of military interest, they were far from being months of idleness or enjoyment. The administrative duties of so large a command drew heavily upon the time and energies of the admiral, and, as has been said, they were not congenial to him. When the Tennessee crossed Dog River Bar, which she did on the 18th of May, Farragut felt that he must be on the spot, in case she attempted to execute her threat of coming out to break up the blockade; but up to that time he was moving actively from point to point of his command, between New Orleans on the one side, and Pensacola, now become his principal base, on the other. From time to time he was off Mobile, and for more than two months preceding the battle of the Bay he lay off the port in all the dreary monotony of blockade service. The clerical labor attaching to the large force and numerous interests entrusted to him was immense. Every mail brought him, of course, numerous communications from the Department. "I received your letter last evening," he writes to a member of his family, "but at the same time received so many from the Department that my eyes were used up before I came to yours, so that mine to you will be short and badly written." A very large part of this correspondence consisted of letters from United States consuls abroad, forwarded through the State Department, giving particulars of vessels fitting or loading for the Confederacy or to break the blockade. "Nearly all my clerical force is broken down," he writes on another occasion. "The fact is, I never saw so much writing; and yet Drayton, who does as much as any of them, says it is all necessary. So I tell them to go on. I do not mind signing my name. Although I write all my own letters, some one has to copy them. My fleet is so large now that it keeps us all at work the whole time."

But while he spoke thus lightly of his own share in these labors, the confinement, the necessary attention to and study of larger details, even while he intrusted the minor to others, and the unavoidable anxieties of a man who had so many important irons in the fire, and at the same time was approaching his sixty-fourth year, told upon him. To this he bore witness when, after the capture of the Mobile forts, the Department desired him to take command of the North Atlantic fleet, with a view to the reduction of Wilmington, North Carolina. "They must think I am made of iron," he wrote home. "I wrote the Secretary a long letter, telling him that my health was not such as to justify my going to a new station to commence new organizations; that I must have rest for my mind and exercise for my body; that I had been down here within two months of five years, out of six, and recently six months on constant blockade off this port, *and my mind on the stretch all the time;* and now to commence a blockade again on the Atlantic coast! Why, even the routine of duty for a fleet of eighty sail of vessels works us all to death; and but that I have the most industrious fleet-captain and secretary, it would never be half done. It is difficult to keep things straight." "I know," he writes on another occasion, "that few men could have gone through what I have in the last three years, and no one ever will know except yourself perhaps. . . . What the fight was to my poor brains, neither you nor any one else will ever be able to comprehend. Six months constantly watching day and night for an enemy; to know him to be as brave, as skilful, and as determined as myself; who was pledged to his Government and the South to drive me away and raise the blockade, and free the Mississippi from our rule. While I was equally pledged to my Government that I would capture or destroy the rebel."

Besides his labors and the official anxieties due to his individual command, he again, as in 1862, felt deeply the misfortunes with which the general campaign of 1864 opened, and especially in the Southwest. There was continually present to the minds of the leaders of the United States forces during the war the apprehension that the constancy of the people might fail; that doubtful issues might lead to a depression that would cause the abandonment of the contest, in which success was nevertheless assured to perseverance and vigor. Grant's memoirs bear continual testimony to the statesmanlike regard he had, in planning his greater military operations, to this important factor in the war, the vacillation under uncertainty of that popular support upon which success depended. The temperament of Farragut reflected readily the ups and downs of the struggle, and was saddened by the weaknesses and inconsistencies of his own side, which he keenly appreciated. "I am *depressed*," he writes, "by the bad news from every direction. The enemy seem to be bending their whole soul and body to the war and whipping us in every direction. What a disgrace that, with their slender means, they should, after three years, contend with us from one end of the country to the other! . . . *I get right sick*, every now and then, at the bad news." "The victory of the Kearsarge over the Alabama," on a more auspicious occasion, "raised me up. I would sooner have fought that fight than any ever fought on the ocean"; and his exultation was the greater that the first lieutenant of the Kearsarge had been with him in the same capacity when the Hartford passed the Mississippi forts.

But, while thus sensitive to the vicissitudes of his country's fortunes, he did not readily entertain the thought of being himself defeated. "As to being prepared for defeat," he wrote before New Orleans, "I certainly am not. Any man who is prepared for defeat would be half defeated before he commenced. I hope for success; shall do all in my power to secure it, and trust to God for the rest." And again: "The officers say I don't believe anything. I certainly believe very little that comes in the shape of reports. They keep everybody stirred up. I mean to be whipped or to whip my enemy, and not to be scared to death." "I hope for the best results," he wrote a week before forcing the passage into Mobile Bay, "as I am always hopeful; put my shoulder to the wheel with my best judgment, and trust to God for the rest"; or, in more homely language: "Everything has a weak spot, and the first thing I try to do is to find out where it is, and pitch into it with the biggest shell or shot that I have, and repeat the dose until it operates." "The Confederates at Fort Morgan are making great preparations to receive us. That concerns me but little"—words used not in a spirit of mere light-heartedness, but because it was a condition he had from the first accepted, and over which he hoped to triumph; for he continues, "I know they will do all in their power to destroy us, and we will reciprocate the compliment. I hope to give them a fair fight if once I get inside. I expect nothing from them but that they will try to blow me up if they can."

Amid such cares and in such a spirit were spent the six months of monotonous outside blockade preceding the great victory that crowned his active career. The only relief to its weariness was a bombardment of Fort Powell, undertaken by the light-draft steamers of the squadron from Mississippi Sound in February, to create a diversion in favor of Sherman's raid from Vicksburg upon Meridian, which was then in progress. The boats could not get nearer to the work than four thousand yards, and even then were aground; so that no very serious effect was produced. A greater and more painful excitement was aroused by the misfortunes of the Red River expedition in April and May. Begun on unsound military principles, but designed politically to assert against French intrigues the claim of the United States to Texas, that ill-omened enterprise culminated in a retreat which well-nigh involved the Mississippi squadron in an overwhelming disaster. The Red River was unusually low for the season, and falling instead of rising. There was not, when the army retired, water enough to enable the gunboats which had ascended the river to repass the rapids at Alexandria. The army could delay but for a limited time, at the end of which, if the boats had not passed, they must be left to their fate. Farragut, who was in New Orleans when the news arrived, wrote bitterly about the blunders made, and was sorely distressed for the issue to the navy. "I have no spirit to write," he says. "I have had such long letters from Porter and Banks, and find things so bad with them that I don't know how to help them. I am afraid Porter, with all his energy, will lose some of his finest vessels. I have just sent him some boats to help him." The boats, however, were saved by the skill and energy of Colonel Joseph Bailey, the chief-of-engineers in Franklin's corps of Banks's army; by whom was thrown across the river a dam, which raised the water on the shoals sufficiently for the boats to cross.

A more pleasant incident occurred to vary the sameness of the blockade days, in the presentation to the admiral, by the Union League Club of New York, of a very handsome sword, with scabbard of massive gold and silver, the hilt set in brilliants. The gift was accompanied by a letter expressive of the givers' appreciation of the brilliant services rendered to the nation, and was a grateful reminder to Farragut, then watching before Mobile for his last grapple with the enemy in his front, that his fellow-countrymen in their homes were not wanting in recognition of the dangers he had incurred, nor of those he was still facing on their behalf.

The time was now close at hand when the weary and anxious waiting, which the admiral afterward so feelingly described, was to be exchanged for the more vigorous action he had so long desired. The co-operation of a division from Canby's army was assured toward the end of July; and at the same time the long-promised, long-delayed monitor ironclads began to arrive. As the want of these and the presence of the enemy's ironclads had been the reasons which, in Farragut's opinion, had made necessary the postponement of the purely naval part of the combined operation, a short description of the vessels which formed so potent an element in his calculations will not be out of place.

The idea of the monitor type of ironclads, which was then the prevalent one in the United States Navy, was brought by John Ericsson from his home in Sweden, where it had been suggested to him by the sight of the rafts with a house upon them crossing the waters with which he was familiar. In its conception, the monitor was simply a round fort, heavily plated with iron, resting upon a raft nearly flush with the water, and provided with the motive power of steam. The forts, or turrets, as they are commonly called, might be one or more in number; and each carried usually two heavy guns, standing side by side and pointing in exactly the same direction, so that if discharged together the projectiles would follow parallel courses. Within the turret the guns could be turned neither to the right nor to the left; if such a change of aim were wished, the turret itself was revolved by steam machinery provided for the purpose. When loading, the port through which the gun was fired was turned away from the enemy; so that if a shot happened to strike at that time it fell on the solid armor. Above the gun-turret there was a second of much smaller diameter, which did not revolve. It was also heavily plated and designed to shelter the commanding officer and those charged with the steering of the ship. So much inconvenience was, however, experienced from smoke and from concussion when these steering turrets were struck, and their dimensions were so contracted, that many captains preferred to remain outside, where they could see better, their orders being transmitted to the helmsmen through the sight-holes pierced in the armor. Of these ironclads, four accompanied Farragut in his attack upon Mobile Bay. Two, the Tecumseh and Manhattan, came from the Atlantic coast, and were sea-going monitors. They had each but one turret, in which they carried two fifteen-inch guns, the heaviest then in use afloat. The other two were river monitors, built at St. Louis for service in the Mississippi. They were consequently of light draught, so much so that to obtain the necessary motive power they each had four screw propellers of small diameter, and they carried four eleven-inch guns in two turrets. Their names were the Winnebago and the Chickasaw. The armor of the two single-turreted monitors was ten inches thick, and that of the river monitors eight and a half inches.

The Tennessee, to which these were to be opposed, was a vessel of different type, and one to which the few ironclads built by the Confederates for the most part conformed—called commonly the broadside ironclad, because the guns, like those of ships-of-war generally, were disposed chiefly along the sides. Her hull was built at Selma, on the Alabama River, and thence towed to Mobile to be plated; it being desirable to take her down the river while as light as possible. She was two hundred and nine feet long and forty-eight feet wide, drawing, as has been said, fourteen feet when loaded. Upon her deck, midway between the bow and the stern, was a house seventy-nine feet long, whose sides and ends sloped at an angle of thirty-four degrees and were covered with iron plating, six inches thick on the forward end and five inches thick on the other end and the sides. With the inclination given, a cannon ball striking would be likely to be turned upward by the iron surface, instead of penetrating. The sloping sides of the house were carried down beyond the point where they met those of the vessel, until two feet below the water. There they turned and struck in at the same angle toward the hull, which they again met six or seven feet under water. Thus was formed all round the ship a knuckle, which, being filled in solid and covered with iron, was a very perfect protection against any but the most powerful ram. The Tennessee herself was fitted with a beak and intended to ram, but, owing to the slender resources of the Confederacy, her engines were too weak to be effective for that purpose. She could only steam six knots. Her battery, however, was well selected and powerful. She carried on each side two six-inch rifles, and at each end one seven-inch rifle—six guns in all. There were, besides the Tennessee, three wooden gunboats, and Farragut was informed that there were also four ironclads; but this, as regards the lower bay at least, was a mistake.

It will be seen from this account, and from the description before given of Mobile Bay, that the advantages of the Tennessee were her great protective strength, a draught which enabled her to choose her own position relatively to the heaviest of the enemy's ships, and the superior range and penetrative power of her guns, being rifles; for while there were cannon of this type in the United States fleet, the great majority of them were smooth bores. The ironclads opposed to her had only smooth-bore guns, incapable of penetrating her side, and therefore only able to reduce her by a continued pounding, which might shake her frame to pieces. The chief defects of the Tennessee as a harbor-defense ship, for which she was mainly intended, were her very inferior speed, and the fact that, by an oversight, her steering chains were left exposed to the enemy's shot. This combination of strong and weak points constituted her tactical qualities, which should have determined the use made of her in the impending battle.

Although the ironclads were, as Farragut esteemed them, the controlling factors in the defense and attack, the Tennessee was by no means the only very formidable obstacle in the way of his success. Except the ironclads, the fleet he carried into Mobile Bay was not substantially stronger than that with which he fought his way up the Mississippi; but since that time the enemy had done much to strengthen the works which he now had to encounter. The number of heavy guns in Fort Morgan bearing upon the channel was thirty-eight. In Fort Jackson, excluding the obsolete caliber of twenty-four pounders, there were twenty-seven, and in St. Philip twenty-one—total, forty-eight; but in caliber and efficiency those of Morgan were distinctly superior to those of the river forts, and it may be considered an advantage that the power was here concentrated in a single work under a single hand. The gunners of Fort Morgan, moreover, had not been exposed to the exhausting harassment of a most efficient bombardment, extending over the six days prior to the final demand upon their energies. They came fresh to their work, and suffered during its continuance from no distraction except that caused by the fire of the fleet itself. While, therefore, Fort Gaines could not be considered to support Morgan by any deterrent or injurious influence upon the United States fleet, the latter work was by itself superior in offensive power to the two Mississippi forts.

To the general defense the Confederates had here brought two other factors, one of a most important and as yet unknown power. As the sand bank extending eastward from Dauphin Island was to some extent passable by light gunboats, a line of piles was driven in the direction of Fort Morgan nearly to the edge of the channel. Where the piles stopped a triple line of torpedoes began, following the same general course, and ending only at a hundred yards from Fort Morgan, where a narrow opening was left for the passage of friendly vessels —blockade runners and others. Had the electrical appliances of the Confederacy been at that time more highly developed, this narrow gap would doubtless also have been filled with mines, whose explosion depended upon operators ashore. As it was, the torpedo system employed at Mobile, with some few possible exceptions, was solely mechanical; the explosion depended upon contact by the passing vessel with the mine. To insure this, the line was triple; those in the second and third rows not being in the alignment of the first, but so placed as to fill the interstices and make almost impracticable the avoidance of all three torpedoes belonging to the same group.

These arrangements were sufficiently well known to Farragut through information brought by refugees or deserters. They—the power of the works, the disposition of the torpedoes, the Tennessee and her companions—constituted the elements of the problem which he had to solve to get his fleet safely past the obstacles into the bay. Although not disposed to lay as much stress as others upon the torpedoes, which were then but an imperfectly developed weapon, prudence dictated to him the necessity of passing between them and the fort; and this was fortunately in accordance with the sound policy which dictates that wooden vessels engaging permanent works, less liable than themselves to penetration, should get as close as possible to the enemy, whose fire they may then beat down by the rapidity of their own. There were certain black buoys floating across the channel, between the piles and Fort Morgan, and it was understood that these marked the position of the torpedoes. The admiral's flag-lieutenant, Lieutenant (now Captain) John C. Watson, had examined these buoys in several nightly reconnaissances; but, although he had not been able to discover any of the mines, the assurances of their existence could not be disregarded. His examination doubtless had some effect upon the admiral's instant determination, in the unforeseen emergency that arose during the action, to pass over the spot where the hidden dangers were said to lie; but in the dispositions for battle the order was given for the fleet to pass eastward of the easternmost buoy, where no torpedoes would be found.

The closeness of this approach, however, and the fact that the line of the channel led in at right angles to the entrance, had the disadvantage of obstructing the fire of the broadside wooden vessels, in which the offensive strength of the fleet, outside the monitors, consisted. The guns of those ships, being disposed along the sides, were for the most part able to bear only upon an enemy abreast of them, with a small additional angle of train toward ahead or astern. It was not, therefore, until nearly up with the fort that these numerous cannon would come into play, and exercise that preponderating effect which had driven off the gunners at Forts St. Philip and Jackson. This inconvenience results from the construction of such ships, and can only be overcome by a movement of the helm causing the ship to diverge from her course; a resort which led a witty Frenchman to say that a ship-of-war so situated is like a shark, that can only bite by turning on its back. The remedy, however applicable under certain circumstances and in the case of a single ship, causes delay, and therefore is worse than the evil for a fleet advancing to the attack of forts, where the object must be to close as rapidly as possible. There are, however, on board such vessels a few guns, mounted forward and called chase guns, which, from the rounding of the bows, bear sooner than the others upon the enemy toward whom they are moving. To support these and concentrate from the earliest moment as effective a fire as possible upon the works, Farragut brought his ironclads inside of the wooden vessels, and abreast the four leaders of that column. The heavy guns of the monitors could fire all around the horizon, from right ahead to right astern; and the disposition had the additional great advantage that, in the critical passage inside the torpedo buoys, these all-important vessels would be on the safer side, the wooden ships interposing between them and the sunken dangers, which threatened an injury far more instantaneous and vital than any to be feared from the enemy's shot and shell.

The position of the ironclads being determined by these considerations, the arrangement of the wooden ships for the attack conformed to the admiral's principle, that the greatest security was to be found in concentrating upon the enemy the heaviest fire attainable from his own guns. As at Port Hudson, a large proportion of the fourteen vessels he purposed to take in with him were of the gunboat class, or a little above it. Resort was accordingly again had to the double column adopted there; the seven ships that had the most powerful batteries forming the right column to engage Fort Morgan. The lighter ones were distributed in the other column, and lashed each to one of the heavier ships, in an order probably designed, though it is not expressly so stated, to make the combined steam power of the several pairs as nearly equal as possible. Among the gunboats there were three that had side-wheel engines, the machinery of which is necessarily more above water, and so more exposed than that of a screw—a condition which, although their batteries were powerful for their tonnage, emphasized the necessity of sheltering them behind other ships during the furious few minutes of passing under the guns of the fort.

The sum of these various considerations thus resulted in the fleet advancing into action in a column of pairs, in which the heaviest ships led in the fighting column. To this the admiral was probably induced by the reflection that the first broadsides are half the battle, and the freshest attack of the enemy should be met by the most vigorous resistance on his own part; but it is open to doubt whether one of these powerful vessels would not have been better placed in the rear. Upon a resolute enemy, the effect of each ship is simply to drive him to cover while she passes, to resume his activity when relieved from the pressure of her fire. The case is not strictly similar to the advance of a column of troops upon a fortified position, where the head does the most of the fighting, and the rear mainly contributes inertia to the movement of the mass. It is at least open to argument that a fire progressively diminishing from van to rear is not, for the passage of permanent works, a disposition as good as a weight of battery somewhat more equally distributed, with, however, a decided preponderance in the van. The last of the ships in this column received a shot in the boiler, which entirely disabled her—an accident that may have been purely fortuitous, and to which any one of her predecessors was in a degree liable, but also possibly due to the greater activity of the enemy when no longer scourged by the more powerful batteries which preceded. She was saved from the more serious results of this disaster, and the squadron spared the necessity of rallying to her support, by the other admirable precautions dictated by Farragut's forethought.

Subjected thus to analysis, there seems much to praise and very little to criticise in the tactical dispositions made by the admiral on this momentous occasion. But the tactical dispositions, though most important, are not the only considerations; it is the part of the commander-in-chief to take advantage of any other circumstances that may make in his favor. Until the forts were passed the character of the bottom left Farragut no choice as to the direction of his attack. There was but one road to take, and the only other question was the order in which to arrange his ships. But there were two conditions not entirely within his control, yet sure to occur in time, which he considered too advantageous to be overlooked. He wanted a flood tide, which would help a crippled vessel past the works; and also a west wind, which would blow the smoke from the scene of battle and upon Fort Morgan, thereby giving to the pilots, upon whom so much depended, and to the gunners of the ships, the advantage of clearer sight. The time of the tide, in most quarters a matter of simple calculation, is in the Gulf often affected by the wind. The wind, on the other hand, in the summer months, blows from the south during the early morning, and then works round to the westward; so that the chances were in favor of his obtaining his wishes.

The dispositions taken by the Confederates to meet the assault which they saw to be impending were more simple; they having but a small mobile force, and their fortifications being tied to their places. A seaport liable to attack is a battle-field, in utilizing whose natural features, so as to present the strongest tactical combination against entrance or subjection by an enemy, the skill of the engineer is shown; but, unlike battle-fields in general, much time and study is allowed to develop his plans. In the case of Mobile Bay, the narrow and direct character of the approach by the main ship channel left little opportunity for skill to display itself. To place at the end of Mobile Point the heaviest fort, enfilading the channel, and to confine the latter to the narrowest bed, compelling the assailant into the most unfavorable route, were measures too obvious to escape the most incapable. To obtain the utmost advantage from this approach of the enemy, the little naval force was advanced from Mobile Point, so as to stretch at right angles across the channel just within the torpedo line. There, without being incommoded by the fire of the fort, or in any way embarrassing it, they secured a clear sweep for their guns, raking their opponents; who, being for the time unable to deviate from their course, could not reply to this galling attack. By gradually retiring, the Confederate gunboats could retain this superiority during the advance of their foes, until the latter reached the wide hole within, where there was room to manœuvre. This position and the subsequent course of action described comprise the tactical management of the Southern vessels during the engagement. It was well devised, and made probably the best use of the advantages of the ground possible to so inferior a force. The Tennessee took position with them, but her after action was different.

As the day of the last and, with the exception of the Essex fight of his boyhood, the most desperate battle of his life drew near, a certain solemnity—one might almost say depression—is perceptible in the home letters of the admiral. Had the action proved fatal to him it could scarcely have failed to attract the attention which is similarly arrested by the chastened tone of Nelson's life and writing immediately before Trafalgar; and although there is certainly none of that outspoken foreboding which marked the last day of the English hero, Farragut's written words are in such apparent contrast to the usual buoyant, confident temper of the man, that they would readily have been construed into one of those presentiments with which military annals abound. "With such a mother," he writes to his son a week before the battle, "you could not fail to have proper sentiments of religion and virtue. I feel that I have done my duty by you both, as far as the weakness of my nature would allow. I have been devoted to you both, and when it pleases God to take me hence I shall feel that I have done my duty. I am not conscious of ever having wronged any one, and have tried to do as much good as I could. Take care of your mother if I should go, and may God bless and preserve you both!" The day before the action he wrote the following letter to his wife, which, as his son remarks in his Life of the admiral, shows that he appreciated the desperate work before him:

"FLAG-SHIP HARTFORD,
"OFF MOBILE, *August* 4, 1864.

"MY DEAREST WIFE: I write and leave this letter for you. I am going into Mobile in the morning, if God is my leader, as I hope he is, and in him I place my trust. If he thinks it is the proper place for me to die, I am ready to submit to his will in that as in all other things. My great mortification is that my vessels, the ironclads, were not ready to have gone in yesterday. The army landed last night, and are in full view of us this morning, and the Tecumseh has not yet arrived from Pensacola.

"God bless and preserve you, my darling, and my dear boy, if anything should happen to me; and may his blessings also rest upon your dear mother, and all your sisters and their children.

"Your devoted and affectionate husband, who never for one moment forgot his love, duty, or fidelity to you, his devoted and best of wives,

"D. G. FARRAGUT."

A more touching and gratifying testimony of unwavering attachment, after more than twenty years of marriage, no wife could desire. It was an attachment also not merely professed in words, but evidenced by the whole course of his life and conduct. Infidelity or neglect of a wife was, in truth, in the estimation of Admiral Farragut, one of the most serious of blots upon a man's character, drawing out always his bitterest condemnation.

A pleasing glimpse is at this same period afforded of his relations to the surviving members of his father's family, who still remained in or near New Orleans, and from whom by the conditions of his profession he had been separated since his childhood. "My dear sister," he writes, "has sent me a Holy Virgin like the one Rose gave me. She said it was blessed by the archbishop, who said I was good to the priests. I only tell you this," adds the admiral dryly, "to show you that they did not succeed in impressing the bishop with the idea that I had robbed the church at Point Coupée." This is not the only mention of his sister during this time, and it is evident that two years' occupation of New Orleans by the Union forces had done much to mollify public sentiment; for immediately after the surrender he had written home, "It is a strange thought that I am here among my relatives, and yet not one has dared to say 'I am happy to see you.'"

On the 8th of July General Canby, accompanied by General Granger, who was to have immediate charge of the land operations against the Mobile forts, had called upon the admiral to make the preliminary arrangements. Somewhat later Canby sent word that he could not spare men enough to invest both Gaines and Morgan at the same time; and at Farragut's suggestion it was then decided to land first upon Dauphin Island, he undertaking to send a gunboat to cover the movement. Granger visited him again on the 1st of August, and as the admiral then had reason to expect the last of his monitors by the 4th, that day was fixed for the attack and landing. Granger was up to time, and his troops were put ashore on the evening of the 3d; but the Tecumseh had not arrived from Pensacola. The other three had been on hand since the 1st, anchored under the shelter of Sand Island, three miles from Fort Morgan.

To Farragut's great mortification he was unable to carry out his part of the programme; but on the evening of the 4th the Tecumseh arrived, together with the Richmond, which had been for a few days at Pensacola preparing for the fight. "I regret to have detained you, admiral," said Craven, the commander of the monitor, "but had it not been for Captain Jenkins (of the Richmond), God knows when I should have been here. When your order came I had not received an ounce of coal." In his report of the battle, Farragut warmly acknowledged the zeal and energy of Jenkins, to which he owed the seasonable arrival of this important re-enforcement. "He takes," he said, "as much interest in the fleet now as formerly when he was my chief-of-staff. He is also commanding officer of the second division of my squadron, and as such has shown ability and the most untiring zeal. . . . I feel I should not be doing my duty did I not call the attention of the Department to an officer who has performed all his various duties with so much zeal and fidelity." Farragut has been charged with failure to notice adequately the services of those under him; but the foregoing words, which are not by any means unparalleled in his dispatches, show that he could praise cordially when he saw fitting occasion.

The night of August 4th was quiet, the sea smooth, with a light air just rippling the surface of the water. At sundown it had been raining hard, but toward midnight cleared off, the weather becoming hot and calm. Later on a light air again sprang up from the southwest. The admiral was not well, and slept restlessly. About three in the morning he called his servant and sent him to find out how the wind was. Learning that it was from the quarter he wished, he said, "Then we will go in in the morning." Between four and five the lighter vessels got under way and went alongside those to which they were to be lashed. When daybreak was reported Farragut was already at breakfast with the captain of the Hartford, Percival Drayton, and the fleet-surgeon, Dr. James C. Palmer, who had left his usual post at the hospital in Pensacola to superintend the care of those wounded in the approaching battle. It was then about half-past five; the couples were all formed, and the admiral, still sipping his tea, said quietly, "Well, Drayton, we might as well get under way." The signal was made and at once acknowledged by the vessels, which had all been awaiting it, and the seamen began to heave round on the cables. The taking their assigned positions in the column by the different pairs consumed some time, during which the flag-ship crossed the bar, at ten minutes past six. At half-past six the column of wooden vessels was formed, and the monitors were standing down from Sand Island into their stations, in gaining which some little further delay was caused. At this time all the ships hoisted the United States flag, not only at the peak where it commonly flies, but at every mast-head as well.

It had been the intention of the admiral to lead the column of wooden vessels with his own ship; but at the earnest request of many officers, who thought the fleet should not incur the greater risk consequent upon having its commander in so exposed a position, he reluctantly consented to waive his purpose, and the Brooklyn was appointed to this post of honor. To this selection contributed also the fact that the Brooklyn had more than the usual number of chase guns, the advantage of which has been explained, and also an arrangement for picking up torpedoes. Bitterly afterward did Farragut regret his yielding on this occasion. "I believe this to be an error," he wrote in his official report of the battle; "for, apart from the fact that exposure is one of the penalties of rank in the navy, it will *always* be the aim of the enemy to destroy the flag-ship, and, as will appear in the sequel, such attempt was very persistently made." "The fact is," he said in one of his letters home, "had I been the obstinate man you sometimes think me, I would have led in the fleet and saved the Tecumseh"—meaning, doubtless, that, by interposing between that important vessel and the buoy which marked the torpedo line, he would have prevented the error which caused her loss. Some notes upon the action found afterward among his papers contain the same opinion, more fully and deliberately expressed. "Allowing the Brooklyn to go ahead was a great error. It lost not only the Tecumseh, but many valuable lives, by keeping us under the fire of the forts for thirty minutes; whereas, had I led, as I intended to do, I would have gone inside the buoys, and all would have followed me." The Hartford took the second place in the column, having secured on her port or off side the side-wheel gunboat Metacomet, Lieutenant-Commander James E. Jouett.

While the monitors were taking their stations, the Tecumseh, which led their column, fired two shots at the fort. At five minutes before seven, the order of battle now being fully formed, the fleet went ahead. Ten minutes later Fort Morgan opened fire upon the Brooklyn, which at once replied with her bow guns, followed very soon by those of the fighting column of wooden ships; a brisk cannonade ensuing between them, the monitors, and the fort. In order to see more clearly, and at the same time to have immediately by him the persons upon whom he most depended for governing the motions of the ship, Farragut had taken his position in the port main-rigging. Here he had near him Captain Jouett, standing on the wheel-house of the Metacomet, and also the pilot, who, as at Port Hudson, had been stationed aloft, on this occasion in the maintop, so as to see well over the smoke. As this increased and rose higher, Farragut went up step by step until he was close under the maintop. Here, without losing touch with Jouett, he was very near the pilot, had the whole scene of battle spread out under his eyes, and at the same time, by bracing himself against the futtock shrouds, was able to use his spy-glass more freely. Captain Drayton, however, being alarmed lest he might be thrown to the deck, directed a seaman to carry a lashing aloft and secure him to the rigging, which the admiral, after a moment's remonstrance, permitted. By such a simple and natural train of causes was Farragut brought to and secured in a position which he, like any other commander-in-chief, had sought merely in order better to see the operations he had to direct; but popular fancy was caught by the circumstance, and to his amusement he found that an admiral lashed to the rigging was invested with a significance equivalent to that of colors nailed to the mast. "The illustrated papers are very amusing," he wrote home. "Leslie has me lashed up to the mast like a culprit, and says, 'It is the way officers will hereafter go into battle, etc.' You understand, I was only standing in the rigging with a rope that dear boy Watson had brought me up," (this was later in the action, when the admiral had shifted his position), "saying that if I would stand there I had better secure myself against falling; and I thanked him for his consideration, and took a turn around and over the shrouds and around my body for fear of being wounded, as shots were flying rather thickly."

Shortly after the monitors and the bow guns of the fleet began firing, the enemy's gunboats and the Tennessee moved out from behind Morgan and took their position enfilading the channel. Twenty minutes later, through the advance of the column, the broadsides of the leading ships began to bear upon the fort; and as these heavy batteries vomited their iron rain the fire of the defense visibly slackened. Amid the scene of uproar and slaughter, in which the petty Confederate flotilla, thanks to its position of vantage, was playing a deadly part quite out of proportion to its actual strength, the Tecumseh alone was silent. After the first two shots fired by her, which were rather the signal of warning than the opening of the battle, she had loaded her two guns with steel shot, backed by the heaviest charge of powder allowed, and, thus prepared, reserved her fire for the Tennessee alone. "I believe," wrote Farragut in a private letter, "that the Tecumseh would have gone up and grappled with and captured the Tennessee. Craven's heart was bent upon it."

The two columns, of ironclads and of wooden vessels lashed together in pairs, were now approaching the line of torpedoes and the narrow entrance through which lay the path of safety; and the broadsides of the heavy sloops which led—the Brooklyn, the Hartford, the Richmond—supported by the less numerous but still powerful batteries following, and by the guns of the turreted ironclads, overbore the fire of the works. All promised fairly, provided the leaders of the two columns pushed rapidly and unhesitatingly in the direction assigned them. But almost at the same moment doubt seized them both, and led to a double disaster. As Craven, leading the monitor column, and then about three hundred yards in advance of the Brooklyn, drew up to the buoy, to the eastward of which he had been directed to go, he saw it so nearly in line with the point beyond that he could not believe it possible to pass. "It is impossible that the admiral means us to go inside that buoy," he said to the pilot; "I can not turn my ship." Just then the Tennessee moved a little ahead, to the westward; and Craven, under the double impulse of his doubt and of his fear lest the hostile ironclad should escape him, changed his course to the left and pushed straight for her, the Tecumseh heading to pass the buoy on the wrong side.

The movement thus indicated, if followed by the succeeding monitors, would throw that column across the path of the wooden ships if the latter endeavored to obey their orders to pass east of the buoy. At the same moment there were seen from the Brooklyn, in the water ahead, certain objects which were taken to be buoys for torpedoes. The ship was at once stopped and backed, coming down upon the Hartford, her next astern, which also stopped, but did not reverse her engines. The Richmond followed the Hartford's movements, and the two ships drifted up with the young flood tide, but with their heads still pointed in the right direction, toward the Brooklyn; the stern of the latter vessel, as she backed, coming up into the wind so that her bows turned toward the fort. Fortunately, the rear ships were some little distance off; but Farragut, ignorant of the cause of the Brooklyn's action, saw his line of battle doubling up and threatened with an almost inextricable confusion, in the most difficult and exposed part of the passage, under a cross-fire from the fort and the enemy's vessels. Immediately upon this frightful perplexity succeeded the great disaster of the day. Craven, pursuing his course across the suspected line of danger, had reached within two hundred yards of the Tennessee, and the crews of both vessels were waiting with tense nerves for the expected collision, when a torpedo exploded under the Tecumseh, then distant a little over five hundred yards from the Hartford. From his elevated post of observation Farragut saw her reel violently from side to side, lurch heavily over, and then go down head foremost, her screw revolving wildly in the air as she disappeared.

It was the supreme moment of his life, in which the scales of his fortunes wavered in the balance. All the long years of preparation, of faithful devotion to obscure duty awaiting the opportunity that might never come—all the success attending the two brief years in which his flag had flown—all the glories of the river fights—on the one side; and on the other, threatening to overbear and wreck all, a danger he could not measure, but whose dire reality had been testified by the catastrophe just befallen under his own eyes. Added to this was the complication in the order of battle ahead of him, produced by the double movements of the Brooklyn and Tecumseh, which no longer allowed him to seize the one open path, follow his own first brave thought, and lead his fleet in person through the narrow way where, if at all, safety lay. The Brooklyn, when she began to back, was on the starboard bow of the flag-ship, distant one or two hundred yards, and falling off to starboard lay directly in the way athwart the channel. The second monitor, Manhattan, of the same class as the Tecumseh, had passed ahead; but the two light-draughts, the Winnebago and Chickasaw, were drawing up abreast of the three ships thus massed together. As they passed, the admiration of the officers of the flag-ship was stirred to see Captain Stevens, of the Winnebago, pacing calmly from turret to turret of his unwieldy vessel, under the full fire of the fort; while of Perkins, in the Chickasaw, the youngest commander in the fleet, and then about twenty-seven years of age, an officer of high position in the flag-ship says, "As he passed the Hartford he was on top of the turret, waving his hat and dancing about with delight and excitement."

But as they went thus gallantly by, the position of these vessels, combined with that of the Brooklyn relatively to the flag-ship, forbade the latter's turning in that direction unless at the risk of adding to a confusion already sufficiently perilous. A signal was made and repeated to the Brooklyn to go ahead; but that vessel gave no sign of moving, her commander being probably perplexed between his orders to pass east of the buoy and the difficulty of doing so, owing to the position into which his ship had now fallen and the situation of the monitors. But to remain thus motionless and undecided, under the fire of the fort with the other ships coming up to swell the size of the target offered to its gunners and to increase the confusion, was out of the question. To advance or to recede seemed alike dangerous. Ahead lay the dreaded line of torpedoes; behind was the possibility of retreat, but beaten, baffled, and disastrous. All depended upon the prompt decision of the admiral. If he failed himself, or if fortune failed him now, his brilliant career of success ended in the gloom of a defeat the degree of which could not be foreseen. In later days, Farragut told that in the confusion of these moments, feeling that all his plans had been thwarted, he was at a loss whether to advance or retreat. In this extremity the devout spirit that ruled his life, and so constantly appears in his correspondence, impelled him to appeal to Heaven for guidance, and he offered up this prayer: "O God, who created man and gave him reason, direct me what to do. Shall I go on?" "And it seemed," said the admiral, "as if in answer a voice commanded, 'Go on!'"

To such a prompting his gallant temper and clear intuitions in all matters relating to war were quick to respond. Personal danger could not deter him; and if it was necessary that some one ship should set the example and force a way through the torpedo line by the sacrifice of herself, he was prepared by all his habits of thought to accept that duty for the vessel bearing his flag. Describing the spirit in which he began an arduous enterprise, after once deciding that it should be undertaken, he said: "I calculate thus: The chances are that I shall lose some of my vessels by torpedoes or the guns of the enemy, but with some of my fleet afloat I shall eventually be successful. I can not lose all. I will attack, regardless of consequences, and never turn back." To a mind thus disciplined and prepared, the unforeseen dilemma presented before the barriers of Mobile Bay caused but a passing perplexity. Like the Puritan soldier who trusted in God and kept his powder dry, Farragut met the overthrow of his carefully arranged plans and the sudden decision thrust upon him with the calm resolution of a man who has counted the cost and is strengthened by a profound dependence upon the will of the Almighty. He resolved to go forward.

The Hartford was now too near the Brooklyn to go clear by a simple movement of her helm. Backing hard, therefore, the wheels of the Metacomet, while turning her own screw ahead, her bows were twisted short round, as in a like strait they had been pointed fair under the batteries of Port Hudson; then, going ahead fast, the two ships passed close under the stern of the Brooklyn and dashed straight at the line of the buoys. As they thus went by the vessel which till then had led, a warning cry came from her that there were torpedoes ahead. "Damn the torpedoes!" shouted the admiral, in the exaltation of his high purpose. "Four bells!* Captain Drayton, go ahead! Jouett, full speed!" The Hartford and her consort crossed the line about five hundred yards from Mobile Point, well to the westward of the buoy and of the spot where the Tecumseh had gone down. As they passed between the buoys, the cases of the torpedoes were heard by many on board knocking against the copper of the bottom, and many of the primers snapped audibly, but no torpedo exploded. The Hartford went safely through, the gates of Mobile Bay were forced, and as Farragut's flag cleared the obstructions his last and hardest battle was virtually won. The Brooklyn got her head round, the Richmond supporting her by a sustained fire from her

---

* The signal in the United States Navy for the engines to be driven at high speed.

heavy broadside; and, after a delay which allowed the flag-ship to gain nearly a mile upon them, the other ships in order followed the Hartford, "believing," wrote the admiral in his dispatch, "that they were going to a noble death with their commander-in-chief."

After the flag-ship had passed the torpedo line the enemy's three gunboats began retreating slowly up the bay, keeping ahead and on her starboard bow, where her guns could not bear while their own raked her. The conditions of the channel did not yet allow her to deviate from her course in order to return their fire. At no period of the battle did the Hartford suffer so much as during the fifteen minutes she had to endure this galling punishment. The Tennessee, being inferior in speed to her consorts as well as to the Hartford, could not accompany this movement; and, moreover, Buchanan, the Confederate admiral, had set his heart upon ramming the vessel that bore the flag of his old friend Farragut. The Tennessee therefore stood toward the Hartford, but failed in her thrust, the Union vessel avoiding it easily with a movement of her helm. The ram then fired two shots at very short range, but singularly enough both missed. "I took no further notice of her," wrote Farragut, "than to return her fire." The Tennessee followed some little distance up the bay, and then, changing her mind, turned toward the column of wooden vessels that was now approaching, with the three monitors covering their right flank and somewhat in the rear; these having delayed to engage the fire of the fort while their more vulnerable companions went by. The Confederate ironclad passed along the column from van to rear, exchanging shots with most of the vessels in it. The Monongahela attempted to ram her, but, being embarrassed by the gunboat lashed alongside, succeeded only in giving a glancing blow; while the Oneida, the ship on the fighting side of the rear couple, already completely disabled in her motive power by a shot through the boiler, received a raking broadside, by which her captain, Mullany, lost an arm.

At the time the Tennessee went about to encounter the remaining vessels of the fleet, which was about eight o'clock, the course of the channel enabled the Hartford to turn sufficiently to bring her broadside to bear on her puny assailants. By the fire she then opened, one, the Gaines, was so much injured as to be with difficulty kept afloat until she could take refuge under Fort Morgan, where she was that night burned by her commander. All three retreated rapidly toward the shoal water on the east side of the bay. Farragut then signaled for the gunboats of his fleet to chase those of the enemy. Jouett, being alongside, received the order by word of mouth, and the admiral often afterward spoke with enthusiasm of the hearty "Ay, ay, sir!" he received in reply, and of the promptness with which the fasts were cut, the men being already by them, hatchet in hand. The Metacomet backed clear at once and started rapidly in pursuit. The gunboats in the rear followed as soon as the signal was made out; but, both from their position and from the inevitable delay in reading signals, they were at a disadvantage. A thick rain squall coming up soon after hid both pursuers and pursued from each other's sight. The Morgan and the Gaines took advantage of it to change their course for Fort Morgan; the third Confederate, the Selma, kept straight on, as did the Metacomet. When the squall cleared, the latter found herself ahead of her chase. One shot was fired, killing the first lieutenant and some of the crew of the Selma, whose flag was then hauled down. The Morgan made good her retreat under the fort, and that night succeeded in escaping up the bay to the city, although she was seen and fired upon by several of Farragut's vessels.

At half-past eight o'clock, three hours after the first signal was made to get under way and an hour and a half after the action began, the flag-ship anchored in the upper part of the deep pocket into which the channel expands after passing the entrance. She was then about four miles from Fort Morgan, and the crew were sent to breakfast. The admiral had come down from his post in the main rigging and was standing on the poop, when Captain Drayton came up to him and said: "What we have done has been well done, sir; but it all counts for nothing so long as the Tennessee is there under the guns of Morgan." "I know it," replied Farragut, "and as soon as the people have had their breakfasts I am going for her." These words were exchanged in the hearing of the first lieutenant of the Hartford, now Rear-Admiral Kimberly, and at present the senior officer upon the active list of the United States Navy. In writing home a few weeks later, the admiral said: "If I had not captured the Tennessee as I did, I should have taken her that night with the monitors, or *tried* it." The latter undoubtedly represents the more deliberate opinion, that would have guided him had Buchanan not played into his hands by attacking the fleet; for if the Tennessee had remained under Morgan and there been sought by the monitors, the fight would have been at such close quarters that in the darkness the fort could scarcely have joined without imminent risk of hurting friend as well as foe.

As it was, the Confederate admiral seems never to have contemplated any more prudent or sagacious course than a single-handed free fight with the fleet. As soon as the Tennessee had passed the rear of the enemy's column, Buchanan said to the captain of the ram: "Follow them up, Johnston; we can't let them off that way." In turning, the Tennessee took much room, appearing from the fleet to have gone back under the guns of Fort Morgan; and the various ships, as they came up, were anchoring near the Hartford, expecting a few quiet hours. They were soon undeceived. The brief conversation above reported between Farragut and his flag-captain had scarcely ended when the ram was seen to be moving out from under the fort. Captain Drayton reported the fact to the admiral, saying that she was going outside to attack the United States vessels still remaining there. "Then," said Farragut, "we must follow him out." The remark indicates an alternative to the course actually adopted by Buchanan, and one whose issue would depend less upon the United States commander-in-chief than upon the conduct of the vessels outside. If these were so imprudent as not to retire, Farragut might have been forced to run twice again the gantlet of Fort Morgan and of the torpedo line—once to protect them, and afterward to regain the position he had just achieved.

It must be admitted that the question before the Confederate admiral, what to do with one unwieldy though powerful vessel opposed to fourteen enemies, was hard to solve; nor did he have, in a precise knowledge of the speed, battery, and other qualities of his opponents, the data needed for an accurate solution. In a general way, however, he must have known that the guns of the United States fleet were mainly smooth-bores, with but moderate penetrative power upon iron-plating such as the Tennessee's; and during the morning's encounter he had acquired experimental knowledge of their impotence against her sides, unless by a continuous pounding such as he was now about to invite. He knew also that several of the hostile vessels were of too heavy draught to take any efficient part, if he refused, as was in his power, to enter the pocket in which they were now anchored; while the general gentle shelving of the bottom enabled a foot's difference in draught to secure a very considerable separation in distance. Every wooden ship was vulnerable to him and impotent against him at the ranges which his rifles permitted him to use.

With the monitors Buchanan had not yet come into collision; but one of the most formidable was sunk, and until he had learned something about their endurance and the power of their guns relatively to those of his own vessel, it would seem that his action, though immediate, should have been only tentative. If it proved on trial that the speed of the Tennessee was greater than that of the monitors, she might yet prove master of the situation. Despite the beak, which her wretched speed and exposed steering chains rendered untrustworthy, her great defensive strength and the fact of carrying rifled guns indicated that long range, and not close quarters, was the first game of the Tennessee. There she could hurt, and she could not be hurt. Had she, for instance, hovered at a distance, firing deliberately at the Union vessels, Farragut must have attacked; and she could then have retired either into shoaler water, retaining her advantage in range, or else under the guns of Morgan, which would have strongly re-enforced her fight. The fact that Farragut, whose instinct for war was commonly accurate, proposed to attack her at close quarters and by night, is the best argument that Buchanan should have sought long range and daylight for his action. As it was, his headlong charge into the Union fleet was a magnificent display of inconsiderate bravery, in which such advantages as he had were recklessly thrown away. Its purpose is not clear. If, as Farragut thought, it was to sink his flag-ship, it can only be replied that an admiral's flag is not a red rag for a bull to charge. Had the Hartford been sunk when the column doubled up an hour or so before, the loss of the leader at so critical a moment might have decided the day; but to sink her in the *mêlée* within would have been a barren, though brilliant, feat of arms.

As soon as it was ascertained that the Tennessee was really coming up to attack, the mess-gear was hurried aside and the orders given to get under way. Some of the fleet had not yet anchored, and the monitors were not yet arrived at the place where the others were gathered. Dr. Palmer, the fleet surgeon, was just leaving the flag-ship in a steam-launch, for the purpose of making a round among the other vessels to see to the condition of their wounded. Farragut called him alongside and directed him to go to the monitors with orders to attack the Tennessee. These Palmer delivered in person to each ironclad. "Happy as my friend Perkins (of the Chickasaw) habitually is," he wrote in his diary, "I thought he would turn a somersault overboard with joy when I told him, 'The admiral wants you to go at once and fight that Tennessee.'" The wooden vessels at the same time were directed to charge the ram, bows on, at full speed, as well as to attack her with their guns.

The monitors being, like the Tennessee herself, very slow, the ramming contest first began. The first to reach the hostile ironclad was the Monongahela, Captain Strong, which struck her squarely amidships on the starboard side, when she was still four hundred yards distant from the body of the fleet. Five minutes later the Lackawanna, Captain Marchand, going at full speed, delivered her blow also at right angles on the port side, abreast the after end of the armored superstructure. As they swung round, both United States vessels fired such guns as would bear, but the shot glanced harmlessly from the armor; nor did the blow of the ships themselves produce any serious injury upon the enemy, although their own stems were crushed in for several feet above and below the water line. Upon them followed the Hartford, approaching, like the Lackawanna, on the port side; but toward her the Tennessee turned, so that the two met nearly, though not exactly, bows on. The Hartford's anchor, which there had not been time to cat, was hanging at the water's edge; it took the brunt of the collision, which doubled it up, and the two antagonists scraped by, their port sides touching. At that close range seven nine-inch guns were discharged against the sloping sides of the ironclad, but without effect. The admiral had clambered again into the rigging, on this occasion into the port mizzen-rigging, whence he watched the effects of this encounter. Both the Lackawanna and the Hartford now made a circuit to get a position whence they could again charge the enemy; but in the midst of their sweep the Lackawanna ran square into the flag-ship, striking near where Farragut stood, and cutting the vessel down to within two feet of the water. The immediate impression among the ship's company was that the injury was fatal; and the general cry that arose, "Save the admiral! Get the admiral on board the Lackawanna!" by its ignoring of their own danger, testified how Farragut's martial and personal qualities had won a way into the affections of his subordinates. With an activity for which he had been remarkable in middle life, and retained even now when in his sixties, the admiral jumped into the chains to ascertain the extent of the injury; then, finding that the ship was in no present danger, he ordered her again to be headed for the Tennessee.

Meanwhile the monitors had come up, and the battle had begun between them and the enemy. One of the Manhattan's fifteen-inch guns had been disabled; and the slow firing of those unwieldy weapons, with the imperfect mechanical appliances then used for loading them, prevented her doing the injury that might have been expected. One shot struck square, breaking through the port side of the armor; but even so the missile itself did not enter the vessel, a strong evidence of the power of the Tennessee to resist a single shot. But she was not equally invulnerable to the sustained and continuous hammering of even lighter projectiles. The Winnebago's turrets, being out of order, could not be turned, and consequently the guns could be brought to bear only by moving the helm; a circumstance which materially reduced her fire. The Chickasaw, however, was in better case. Lieutenant-Commander Perkins got her into position under the stern of the Tennessee just after the latter's collision with the Hartford; and there he stuck to the end, never over fifty yards distant, and keeping up a steady rapping of eleven-inch shot upon the fabric which they could not at once penetrate, but which they visibly shook. Fifty-two of these projectiles were fired from the Chickasaw in the short half-hour of her attack. The exposed rudder-chains were shot away, and at nearly the same time the smoke-stack came down. Admiral Buchanan was wounded by an iron splinter, which broke his leg and otherwise injured it to such an extent that the limb was with difficulty saved. He turned over the command to Captain Johnston, who stood the pounding for twenty minutes longer and then reported to his superior that the ship was helpless, could not be steered, and that for half an hour he had not been able to bring a gun to bear. "Well," replied Buchanan, "if you can not do them any further damage you had better surrender."

The Tennessee's flag had been several times shot away, and was now flying from a boat-hook. Not being very conspicuous, its removal was not immediately noticed, and Johnston had to show a white flag to put a stop to the firing. "She was at this time sore beset," said Farragut in his dispatch to the Navy Department; "the Chickasaw was pounding away at her stern, the Ossipee was approaching her at full speed, and the Monongahela, Lackawanna, and Hartford were bearing down upon her, determined upon her destruction. Her smoke-stack had been shot away, her steering chains were gone, compelling a resort to her relieving tackles, and several of her port shutters were jammed. Indeed, from the time the Hartford struck her until her surrender she never fired a gun." No stronger evidence can be offered than this last sentence, which Johnston's account corroborates, of how completely Buchanan miscalculated, or disregarded, the capabilities of the important vessel he controlled. Great as was her power to resist a single shot, or the end-on charge of a heavy vessel, when she surrendered nearly all the plating on the after side of the casemate was found to be started, and the after gun-carriage was disabled; there being distinct marks of nine eleven-inch solid shot having struck within a few square feet of that port. Three of her port shutters also were so damaged that their guns could not be fired.

Thus ended the great battle of Mobile Bay, the crowning achievement of Farragut's naval career; "one of the hardest-earned victories of my life," to quote his own words, "and the most desperate battle I ever fought since the days of the old Essex." "You may pass through a long career and see many an action," he remarked to one of the junior officers of the Hartford, in the interval between first anchoring and the conflict with the Tennessee, "without seeing as much bloodshed as you have this day witnessed." The loss of the flag-ship herself had been twenty-five killed and twenty-eight wounded out of a ship's company of some three hundred souls. The Brooklyn, a ship of the same force, had almost exactly the same number of casualties—eleven killed and forty-three wounded. Contrasting the equal suffering of the latter—delayed so long under the numerous guns of the fort, but supported by the fire of the other vessels—with that of the flag-ship, inflicted by the batteries of the enemy's gun-boats, few in number, but worked for the time with impunity, we find an excellent illustration of Farragut's oft-repeated maxim, that "to hurt your enemy is the best way to keep him from hurting you." The total loss of the United States fleet in the battle was three hundred and thirty-five; of whom one hundred and thirteen were at the bottom of the bay, coffined in the iron hull of the Tecumseh.

Not quite three hours elapsed from the time that Morgan fired its first gun to the moment when the Tennessee hauled down her flag and confessed the United States fleet mistress of the bay. The forts still stood with the Confederate flag flying from them in defiance; and it is reported that the commander of Morgan retorted to a summons to surrender, that he looked upon Farragut's fleet as practically prisoners in a port whose keys he held. If so, it was the high-hearted resolve of a man determined to hold his charge to the last, and not the sober conviction of a soldier, that spoke. Like the river forts when Farragut's fleet forced its way past and stood between them and their base of supplies, the defenses of Mobile were isolated by the results of the morning's fight, and their fall became but a question of time. There was no mutiny of the garrison, as on the former occasion, for the stern experience of war had better taught the men the business of a soldier; but it was at once practicable here to begin siege operations, which in the river would perhaps have been for a time postponed, owing to the overflowed state of the country. The preparations for these were pushed with vigor, and the navy also took a hand against the works. Four hours after the surrender of the Tennessee, the Chickasaw weighed her anchor and steamed down toward Grant's Pass to shell Fort Powell. Built to resist an attack from Mississippi Sound, the work was weak in the direction of the bay. "The iron-clad's fire," reported the officer in command, "made it impossible to man the two guns in the rear, and I made no attempt to do so." That night the fort was evacuated and blown up. The following day the Chickasaw threw some shells into Fort Gaines, in consequence of which, and of the progress made by General Granger in his approaches, that work was surrendered on the 7th of August. Morgan still standing out, the army was transferred from Dauphin Island to Mobile Point, batteries were constructed, and on the 17th a siege train from New Orleans was landed. On the 22d, at daylight, the siege guns, the three monitors, the captured Tennessee, and the ships, both outside and inside the bay, opened together. The following day Fort Morgan capitulated.

A gratifying feature in these operations, as well as in all Farragut's official association with the army, was the cordial good feeling and co-operation which existed between the two services, and which were equally manifested in the upper Mississippi between Grant and Porter. General Butler, Farragut's first colleague in the Gulf and at New Orleans, but who had long since left the department, wrote him a most enthusiastic letter of congratulation upon receiving the news of the battle of Mobile Bay; and General Granger, in concluding his report of the siege operations against Gaines and Morgan, said: "I am pleased to record the perfect harmony existing between these two branches of the service. For my own part, I can not sufficiently acknowledge the assistance rendered by the fleet and the admiral in command in transporting and disembarking the troops, guns, and materials employed by me in the operations. In brief, during all our relations, the officers of the fleet, with their distinguished commander, displayed in a high degree those qualities which mark their gallant service." To the officers of the navy the testimonies thus given can not but be most grateful; not merely as acknowledgments of the important part played by a service whose work is too often ignored by historians, but chiefly as giving an added lustre to the brilliant reputation of its two most distinguished representatives, who successively filled the high position of admiral of the navy.

After the capitulation of the forts, Admiral Farragut remained in Mobile Bay until the following November. The lower bay was cleared of torpedoes and reconnoissances made toward Mobile; but he wrote adversely to any attempt against the city, now that it was sealed as a port to blockade runners. "It would be an elephant," he wrote, "and take an army to hold it. And besides, all the traitors and rascally speculators would flock to that city and pour into the Confederacy the wealth of New York." He confesses also his dislike to operations in very shoal water. "I am in no way diffident about going anywhere in the Hartford, but when I have to leave her and take to a craft drawing six feet of water I feel badly."

The admiral's health was now suffering much from the combined effects of his labors, his anxieties, and the climate. "I am as well as a man can be who can neither sit, walk, nor stand five minutes at a time on account of Job's comforters. But, thank God (I have so much to be thankful for that I am thanking him all the time), I am otherwise in pretty good condition." Despite this brave effort at cheerfulness, his letters from time to time began to show symptoms of depression, and he longed for rest. "This is the last of my work," he said, "and I expect a little respite." His enfeebled condition drew the attention and excited the alarm of those about him. "I was talking to the admiral to-day," wrote Perkins, of the Chickasaw, the day after Morgan surrendered, "when all at once he fainted away. He is not very well and is all tired out. It gave me quite a shock, and shows how exhausted he is, and his health is not very good, any way. He is a mighty fine old fellow." Captain Drayton also wrote home to his family that, if the admiral remained longer in the Gulf, he feared for the consequences.

Under these circumstances an order from the Navy Department, dated the 5th of September, assigning him to the command of the Fort Fisher expedition, greatly upset him. He had about a week before written to the Secretary to say that his strength was almost exhausted. "I am willing," he concluded, "to do the bidding of the department as long as I am able to the best of my abilities. I fear, however, that my health is giving way. I have now been down in the Gulf five years out of six, with the exception of the short time at home last fall; the last six months have been a severe drag upon me, and I want rest, if it is to be had."

To so reasonable a request, after such distinguished and valuable service, the department could not have closed its ears had it been so disposed. Farragut was authorized to leave his squadron in charge of Commodore James S. Palmer, a very gallant and efficient officer, and to come north in the Hartford. On the 30th of November, 1864, he sailed from Pensacola, and on the 12th of December the flag-ship again anchored in New York Harbor.

CHAPTER XI.

LATER YEARS AND DEATH.

1864–1870.

WITH the strong national and patriotic feeling that had been aroused throughout the Northern States by the war of secession, Farragut had no cause to complain of ingratitude or indifference on the part either of the Government or of his fellow-countrymen. As the flag-ship entered the Narrows, on his final return from the Gulf, she was met by a representative committee from the city officials and citizens of New York. Enthusiastic crowds greeted him as he landed at the Battery, and a reception given him the same afternoon at the Custom House was thronged by the leading men of the city. This eager manifestation of good-will and admiration was followed, a few days later, by a flattering request that the admiral would honor the city by taking up his abode in it and becoming thenceforth one of its citizens. After reciting the deeds which had won for him universal applause and thankfulness, the committee said: "The citizens of New York can offer no tribute equal to your claim on their gratitude and affection. Their earnest desire is to receive you as one of their number, and to be permitted, as fellow-citizens, to share in the renown you will bring to the metropolitan city. This desire is felt in common by the whole community."

This graceful tribute of words was accompanied by the gift of fifty thousand dollars, to facilitate Farragut's complying with the request. The letter was addressed to Vice-Admiral Farragut; the United States Government, not to be behindhand in acknowledging its debt to its most distinguished seaman, having created for him that grade soon after his arrival. The bill for the purpose was introduced on the 22d of December, 1864, immediately passed by both houses, and became law by the President's signature the following day. Farragut's nomination and confirmation followed of course and at once; so that his promotion came to him in the Christmas holidays. The admiral gratefully acknowledged the warm welcome of the New Yorkers, while modestly disavowing, as far as he could, his claim to extraordinary merit in the brilliant services which he asserted were but the performance of his duty; and he thankfully accepted, as the spontaneous offering of his fellow-countrymen, the recompense which in older countries is the usual reward of distinguished military success, but conferred there through the formal medium of the central government.

Toward the end of January, 1865, the Confederate vessels in the James made an attempt to descend the river, destroy the pontoon bridges of the United States armies, and cut off both the Army of the James and that of the Potomac from their base of supplies at City Point. Rear-Admiral David D. Porter, who then commanded the North Atlantic Squadron, was fully occupied at the time with the bombardment of Fort Fisher and capture of Wilmington, North Carolina; and as the hostile attempt threatened a very serious annoyance to the communications of the army, Farragut, who was then in Washington, was ordered to proceed to the spot. He accordingly hoisted his flag on a small steamer and ran down to the James; but, finding upon his arrival that the enemy had been repulsed, and satisfactory measures taken to prevent a renewal of the effort, he returned to Washington. This slight episode concluded his active service in the war.

When Richmond was evacuated on the 2d of April, 1865, Farragut was among the first to visit the fallen capital of the Confederacy. From there a few days later he visited his old home in Norfolk. Many of his former friends still retained strong feelings of resentment against him, as a Southern man who had taken arms against the South. The impression had obtained among some that, though leaving his old home, he would remain neutral; and it was even reported that he had said he would take no part in the war. That Farragut never passed through that phase of feeling, in the struggle between life-long affections and the sense of duty, would be too much to affirm; but it was a position in which a man of his decided and positive character could not have stopped when civil strife was upon the land. It was inconsistent with his general habits of thought; and it is evident that, before leaving Norfolk, his convictions on the particular crisis had already left far behind any such temporary halting place between two opinions. When he justified to his excited neighbors President Lincoln's call for troops, on the ground that the United States Government could do no less, when its arsenals and navy yards were seized and its flag fired upon, it is inconceivable that the man who then had such courage of his opinions entertained any further doubt as to his future course; though it may well be that he did not imperil his personal liberty and safety by any irritating avowal of his purpose. In a reception given to him, when he thus revisited the place which should no longer be his home, he recalled those days and said: "I was told by a brother officer that the State had seceded, and that I must either resign and turn traitor to the Government which had supported me from my childhood, or I must leave this place. Thank God! I was not long in making my decision. I have spent half my life in revolutionary countries, and I know the horrors of civil war, and I told the people what I had seen and what they would experience. They laughed at me, and called me 'granny' and 'croaker'; and I said: 'I can not live here, and will seek some other place where I can live, and on two hours' notice.' I suppose they said I left my country for my country's good, and thank God I did! I was unwilling to believe that this difficulty would not have been settled; but it was all in vain, and as every man must do in a revolution, as he puts his foot down, so it marks his life."

In the summer of 1865, following the close of the war, Farragut visited several of the New England cities, receiving everywhere marks of love and admiration similar to those tendered to him in New York; but his life for the next two years was passed in comparative retirement, seeking the reestablishment of his health, which had been severely shaken by the exposures and anxieties of the war. Though for the most part unassigned to any special duties, the winding up of the affairs of the West Gulf Squadron fully occupied his time.

On the 25th of July, 1866, Congress passed a law creating the grade of admiral in the United States Navy, a position which was of course given at once to Farragut, and has been held by but one other—the late Admiral David D. Porter. The following year he was appointed to command the European Squadron, his flag being hoisted on board the steam frigate Franklin on the 17th of June, 1867. Without any request, and indeed without any expectation, on his part, the Government sent the admiral permission for Mrs. Farragut and a kinswoman to accompany him during the cruise. On the 28th of June the ship sailed from New York,* and on the 14th of July anchored in Cherbourg, France.

After passing a fortnight there, during which the admiral visited Paris and dined with the Emperor, the Franklin sailed for the Baltic, where the months of August and September were passed in visiting the ports of Russia, Sweden, and Denmark. Everywhere Farragut was received with the enthusiasm and distinguished consideration that were aroused among naval officers, by the presence of the man who had bestowed upon their profession a lustre unequaled by any other deeds of that generation. Toward the end of September he arrived in England, where a month was spent in a similar gratifying manner; attentions being lavished upon him by men

* Before the admiral's departure from New York he gave a grand reception on board the flag-ship, which was attended by the President and his Cabinet and by many of the most prominent people of the Metropolis, including several hundred ladies.—Editor.

not only of his own calling, but of all positions. Here, as in the Baltic, every opportunity was given Farragut for visiting all objects of general interest, as well as for examining the professional improvements of the day.

From England the Franklin went to the Mediterranean, which Farragut had not seen since the flying trip made by the Brandywine in the winter of 1825, after landing Lafayette in France. Between October, 1867, and April, 1868, were visited Lisbon, Gibraltar, and several ports of the western Mediterranean belonging to Spain, France, and Italy. Everywhere the same cordial welcome was extended, and the most ample facilities enjoyed for seeing thoroughly the points of interest in which the Mediterranean abounds. At Nice he was the object of especial attentions from the numerous Americans who throng that attractive winter resort; and while at Naples a special excavation was made at Pompeii for his benefit. Nowhere, however, did he have a more elaborate and, from the professional point of view, more interesting reception than in Malta, the great British stronghold in the central Mediterranean; where the Mediterranean fleet, then on the point of sailing for the Levant, was detained especially to meet him.

The incidents of this cruise which most nearly touch Farragut himself, and have the greatest interest for his biographer, occurred in the island of Minorca, where his family originated. Over forty years had passed since, as midshipman and lieutenant, he had wintered at Port Mahon. During those early visits he had received messages from persons living in the interior of the island who claimed relationship; but with boyish indifference he had not responded to any of these advances. Since that time he had become imbued with the interest men commonly feel, in advancing years, in collecting all traces of family history which they can find; especially when, as in his case, they have been early and completely separated from the home of their childhood and of their race. The late George Tickner had sent him an old Spanish book, the poems of Mossen Jaime Febrer, in which he read the account of his earliest celebrated ancestor, Pedro Ferragut. Among several escutcheons of the family that have been preserved, bearing diverse ecclesiastical and military emblems indicative of the individual's profession, all contain the common distinguishing device of a horseshoe; and this the admiral, moved by the feeling of kinship, had adopted for his plate. Drawn by these ties of blood and by curiosity, it was a matter of course that Farragut should visit the famous harbor for which British, French, and Spaniards had battled, and which lay within the limits of his command. The renown of his achievements had carried his name to Ciudadela, the remote inland city where his father was born over a century before; and the quiet islanders, who had exulted in the fame of one sprung from their race, were ready to greet him and claim him as their own. In response to an invitation given by them, the admiral, in December, 1867, paid a visit to Ciudadela, of which the following account is given by his secretary, Mr. Montgomery, who accompanied him on the trip:

"The day after Christmas had been designated by the admiral for his promised visit to Ciudadela, in response to the cordial invitation of the authorities and people of that city. The news of this tour of pleasure had spread rapidly to all parts of the island, and occasioned a general rest from labor and a popular concentration upon the lines of travel. At the towns of Alayor and Mercadal flocks of people of both sexes had assembled on the roadside to unite with the authorities in tendering our naval chieftain a cordial welcome, and in expressing their delight at his advent.

"Although unable to accept the offers of hospitality which even in these unpretending villages were showered upon him, the admiral heartily acknowledged the gratification he felt at their demonstrations of personal regard, and, passing along the excited lines, he underwent a siege of hand-shaking. At these points and elsewhere along the route soldiers had been stationed to pay him proper honors, and to tender him any assistance he might require throughout his journey.

"On his arrival within four miles of Ciudadela he was formally received by the Alcalde; and a large committee, comprising many prominent citizens, tendered the hospitalities of the city, and cordially welcomed him as its guest. After a brief interchange of courtesies, he was transferred to a very handsome barouche, and conducted forward in the van of a quite formidable-looking procession, demonstrations of every kind increasing as he approached this ancient capital of Minorca, the present residence of many of those who prefer the quiet seclusion of their island home to the more dazzling notoriety incident to many of the older and gayer provinces of the mainland. Outside the walls of the city his appearance was no sooner heralded than masses of people of every age, sex, and condition rushed forward to greet him, filling the air with cheers and acclamations. As he passed the gates of the city, the walls, house-tops, and balconies were crowded with anxious spectators, uniting demonstrations of welcome with equally expressive shouts from the swaying multitude who had taken possession of the principal thoroughfares. One old man of threescore years and ten, with tears streaming down his weather-beaten face, stamped sincerity itself upon the nature of the welcome by shouting aloud: 'He is ours! he is ours! but I shall never see him more.'

"The avenue leading to the residence of Señor Don Gabriel Squella, which had been kindly placed by that gentleman at the disposal of the admiral and his suite, was literally blocked with people, and the excitement rose rapidly to fever heat as the head of the column appeared in view endeavoring to make a breach in a body absolutely closed in mass. It was with no little difficulty that the procession forced a passage; and although policemen did their utmost, and jostled, and crowded, and threatened, accompanying their language with all the vocabulary of Spanish expletives, it was found necessary to disembark at some distance from the hospitable mansion and trust to the humanity of our entertainers to afford an entrance on foot. But the temporary concealment of the admiral within the delightful headquarters which had been assigned him seemed to be the signal for a renewed outburst, which brought him to the balcony, upon which he stood bowing his thanks and acknowledging in every possible way his heartfelt appreciation of the cordial welcome extended him, until it appeared that there was no prospect of a cessation of hostilities, when, for the first time in his life, he was persuaded to retreat in the face of superior numbers.

"The excitement continued unabated, however, throughout the entire evening, and it was not until near midnight that the crowd slowly dispersed, and the peaceful little city of Ciudadela resumed its wonted quiet, and its order-loving citizens, unaccustomed to all such sounds of revelry by night, retired to their own little homesteads.

"During this time a fine band of music was stationed in the capacious vestibule on the first floor of Señor Squella's mansion, and almost all the prominent citizens of the place, with their families, called to pay their respects to the city's guest, making the scene of excitement within as pleasant as that without was tumultuous.

"On the following morning enthusiasm arose with the sun, once more took firm possession of the street fronting the headquarters of the admiral, and there kept anxious watch. I am confident that, had there been an election that day for Governor of the Balearic Islands, or for King of Spain itself, the admiral would have been chosen without opposition.

"At an early hour, accompanied by his entire suite, all surrounded and followed by an admiring and excited throng, he was escorted by the committee and other citizens to all the places of interest in and about the city, and finally to the cathedral, in which he had scarcely been seated before it was literally packed in every part by people, their hundreds of eyes being riveted upon the pleasant countenance of the unappalled admiral, who withstood the onslaught with as much *sang froid* as if accustomed to such trying ordeals.

"Soon after, the great organ pealed forth our own national melodies, recalling our far-off land even to those whose knowledge of its power and glory was limited to its history, and the sparse information derived from the few Americans who have visited this secluded city."

After leaving the Mediterranean in April, 1868, the Franklin went to Holland and Belgium, and thence made a second visit to England, in the course of which Farragut was presented to Queen Victoria, and visited Scotland and the north of England. In July he returned to the Mediterranean and made a round of the Levant, visiting Constantinople; a special indulgence to anchor before the city being accorded to the ship bearing the flag of an admiral, whose exceptional achievements made it unlikely that the privilege would shortly be construed into a precedent. After a short stay in Athens, and a run up to Trieste at the head of the Adriatic, the Franklin returned to Gibraltar, and thence sailed for New York, which she reached on the 10th of November, 1868; thus concluding a cruise which, from the beginning to the end, had resembled a triumphal progress in the enthusiastic recognition everywhere extended to the hero, whose battle-won blue flag she carried at her main.

Less than two years of life remained to Admiral Farragut when he returned from the Mediterranean. The following summer of 1869 he visited the California coast, where he had not been since he gave up the command of the Mare Island Navy Yard in 1858. The welcome here accorded him was as hearty as that extended in foreign countries, and mingled with the admiration due to the conquering admiral was the recollection of warm mutual affection and esteem engendered by four years of close intercourse. Returning from San Francisco to the East, Farragut was seized at Chicago with a violent illness, in which the heart was affected. For some days his life was despaired of; and although by careful nursing he recovered so as to resume his journey, it is doubtful whether he ever regained the ground then lost. Several severe attacks followed this one; and although he rallied with extraordinary rapidity, thanks to a vigorous constitution, it was apparent that his health was failing. A few months later, in the middle of winter, he consented to take charge of the naval ceremonies in honor of the remains of Mr. George Peabody, whose body had been brought to the United States in the British ship-of-war Monarch, in recognition of his benevolence to the poor of London. It was his last official duty, and the exposure attendant upon funeral ceremonies in that bleak season was much to be deprecated in a man of his years and failing vigor.

The following summer the Navy Department placed at his disposal the dispatch steamer Tallapoosa, which took him and his family to Portsmouth, New Hampshire; where he became the guest of the late Rear-Admiral Pennock, then commandant of the Navy Yard at that place and a connection by marriage of Mrs. Farragut. It was his last sea voyage, and he appeared to have a presentiment that it was so; for as the ship drew near the yard he arose from his sick bed at the sound of the salute being fired in his honor, dressed himself in full uniform, and went on deck. Looking up with a sad smile at his flag flying from the mast-head, he said: "It would be well if I died *now*, in harness." Shortly after his arrival, an old sailor who had charge of the sloop-of-war Dale, then lying dismantled at the wharf, met there the admiral, who had wandered on board. He looked about the ship and, as he left her to go ashore, said: "This is the last time I shall ever tread the deck of a man-of-war." This prediction proved true. He passed quietly away at the commandant's house, on the 14th of August, 1870, aged sixty-nine years; surrounded by his family and loving friends, including many of his old companions in arms. The body was laid temporarily in Portsmouth, the naval officers and citizens of the place uniting to pay every respect to his memory.

In September the Navy Department sent the steam frigate Guerrière to bring the admiral's body to New York. This ship running aground on Nantucket Shoal, the remains were transferred to another vessel and so conveyed to the city. The final and public funeral ceremonies were held on the 30th of September; the day being observed as one of general mourning, the city edifices draped, bells tolled, and minute guns fired. In the procession was General Grant, then President of the United States, with the members of his Cabinet, many military and naval officers, ten thousand soldiers, and a large number of societies. By these the coffin of the admiral was escorted to the railroad station, whence it was transported to Woodlawn Cemetery, in Westchester County, where the body now lies.

To his memory the United States Government has erected a colossal bronze statue in the national capital, in Farragut Square, the work of Miss Vinnie Ream. A committee of New York citizens have placed a similar memorial, by Mr. St. Gauden, at the northwest corner of Madison Square in that city. There is also a mural tablet, with a likeness of the admiral, in the Protestant Episcopal Church of the Incarnation; of which he was a communicant after taking up his residence in New York.

CHAPTER XII.

THE CHARACTER OF ADMIRAL FARRAGUT.

THE brilliant and victorious career which has secured for Farragut a leading place among the successful naval commanders of all time was of brief duration, and began at an age when men generally are thinking rather of relaxing their efforts than of undertaking new and extraordinary labors. The two great leaders of the United States armies during the civil war—Grant and Sherman—were not over forty-five when the return of peace released them from their cares; while Nelson and Napoleon were but a year older than these when Trafalgar and Waterloo terminated their long careers. Farragut was nearly sixty-one at the time of passing the Mississippi forts, and his command of the Western Gulf Squadron lasted not quite three years, or rather less than the ordinary duration of a naval cruise in times of peace. Though not unprecedented, the display of activity and of sustained energy made by him at such an advanced period of life is unusual; and the severity of the strain upon the mental and physical powers at that age is evidenced by the prostration of Farragut himself, a man of exceptional vigor of body and of a mental tone which did not increase his burdens by an imaginative exaggeration of difficulties. He never committed the error, against which Napoleon cautioned his generals, "*de se faire un tableau.*" On the other hand, the study of his operations shows that, while always sanguine and ready to take great risks for the sake of accomplishing a great result, he had a clear appreciation of the conditions necessary to success and did not confound the impracticable with the merely hazardous. Of this, his reluctance to ascend the Mississippi in 1862, and his insistence in 1864 upon the necessity of ironclads, despite his instinctive dislike to that class of vessel, before undertaking the entrance to Mobile Bay, are conspicuous illustrations; and must be carefully kept in view by any one desirous of adequately appreciating his military character.

As in the case of Nelson, there is a disposition to attribute Farragut's successes simply to dash—to going straight at the enemy regardless of method and of consequences. In the case of the great British admiral the tendency of this view, which has been reproduced in successive biographies down to the latest, is to sink one of the first of naval commanders beneath the level of the pugilist, who in his fighting does not disdain science, to that of the game-cock; and it is doubtless to be attributed to the emphasis he himself laid upon that direct, rapid, and vigorous action without which no military operations, however wisely planned, can succeed. In the want of this, rather than of great professional acquirements, will be most frequently found the difference between the successful and the unsuccessful general; and consequently Nelson, who had seen so much of failure arising from slowness and over-caution, placed, and rightly placed, more stress upon vigor and rapidity, in which most are found deficient, than upon the methods which many understand, however ill they may apply them. Like the distinguished Frenchman, Suffren, who is said to have stigmatized tactics as "the veil of timidity," yet illustrated in his headlong dashes the leading principles of all sound tactics, Nelson carefully planned the chief outlines of operations, in the execution of which he manifested the extremes of daring and of unyielding firmness. There was in him no failure to comprehend that right direction, as well as vigor and weight, is necessary to a blow that would tell; but experience had taught him that the average man wants to be much too sure of success before venturing to move, and hence the insistence upon that one among the features of his military character which to the superficial observer has gradually obscured all others. Vigor even to desperateness of action both Nelson and Farragut on occasion showed—recklessness never. Neither fought as one who beateth the air; and while for neither can be claimed an entire exemption from mistakes, the great outlines of their action can safely challenge hostile criticism.

While, however, both in their respective spheres illustrated the great leading principles of war, the circumstances under which they were called to practice them were too diverse to permit any close comparison, or parallel, to be instituted between their actions. Nelson, for the most part, shone upon the battle-field—by his tactical combinations, by the rapidity and boldness with which he carried out plans previously laid, or, on occasion, by the astonishing *coup d'œil* and daring with which, in unforeseen crises, he snatched and secured escaping victory. Farragut in actual battle showed that careful adaptation of means to ends which has a just claim to be considered tactical science; but his great merit was in the clearness with which he recognized the decisive point of a campaign, or of a particular operation, and threw upon it the force under his direction. Nelson acted chiefly against ships, against forces of a type essentially the same as his own, and accessible in all parts to his attack, because belonging to the same element; he might therefore hope to overcome them by the superior quality of his crews or by his better tactical dispositions. Farragut contended with fortifications, whose military powers, offensive and defensive, were essentially different from those of a fleet. Their endurance so greatly exceeded that of his ships as to exclude any hope of reducing them by direct attack; and their advantages of position, deliberately chosen and difficult of approach, could not be outweighed by any tactical arrangement open to him to adopt. He was therefore compelled to seek their fall by indirect means, by turning and isolating them, by acting against their communications—a conception not tactical, but strategic.

It is not meant to imply that the military talents of either admiral were confined to the particular field ascribed to him, but simply that in general they were led by circumstances to illustrate that chiefly. Nelson in his fine campaign in the Baltic evinced his profound intuitions in the science of strategy; and Farragut, as has been said, showed no mean tactical ability in the provisions made for his several battles. The dispositions to be adopted were with him the subject of very careful consideration; and before Mobile he spent hours with his flag lieutenant studying, by the aid of little wooden models, the different positions in which the ships might be placed. Afterward he had the squadron get under way several times to practice keeping close order, and changing formation and course.

Like all men who have achieved eminence, the secret of Admiral Farragut's success is to be found in natural aptitudes carefully improved, and in a corresponding opportunity for action. How much he was indebted to the latter, is evident from the fact that he had passed his sixtieth year before his great qualities were manifested to the world. He was fortunate also, as was Nelson, in the conditions which he was called to meet. Great as were the difficulties confronting each, and brilliantly as they rose to the demand made upon their energies, it may safely be said that more perfect preparation upon the part of their enemies would either have detracted from the completeness of their victories; or else, by imposing greater deliberation and more methodical execution, would have robbed their exploits of that thunderbolt character which imparts such dramatic brilliancy to the Nile and Trafalgar, to New Orleans and Mobile Bay. A modern torpedo line would not leave the gap by which Farragut first meant to profit, nor would it be crossed with the impunity he found; nor could Nelson in his day, without courting destruction, have used against a thoroughly efficient enemy the tactics that admirably suited the conditions in Aboukir Bay and off Cape Trafalgar. But these considerations do not diminish the credit of either admiral, though they help to explain the fullness of their success, and justify proceedings which under different circumstances would be unjustifiable. Rather, it may be said that, in the adaptation of their measures to the conditions opposed to them, what would otherwise invite condemnation as rashness, demands recognition as genius.

For Farragut had a natural genius for war, to which scarcely any opening had been offered before the unexpected calamity of the great civil strife burst upon the country. In estimating his military character and rightly apportioning the credit due to his great achievements, much stress must be laid upon the constant effort for professional improvement made by him from his early life. "Without the opportunity and the environment which enabled him to develop himself," writes one who knew him for over forty years, "Farragut might have gone to his rest comparatively unknown; yet among his comrades and contemporaries in the navy he would have been recognized as no ordinary man, no merely routine naval officer, who kept his watch and passed through life as easily as he could." "He told me," writes another, who first met him after his flag was flying, "that there are comparatively few men from whom one could not learn something, and that a naval officer should always be adding to his knowledge; it might enable him to be more useful some day; that it was hard to say what a naval officer might not have to do." Even after the war, when his reputation was at its height, in visiting European ports he never for a moment lost sight of this duty of professional acquirement. Not a harbor was visited that he did not observe critically its chances for defense by sea or land. "Who knows," said he, "but that my services may be needed here some day?" "Ah, Mr. Tucker," said Earl St. Vincent to his secretary when planning an attack upon Brest, "had Captain Jervis * surveyed Brest when he visited it in 1774, in 1800 Lord St. Vincent would not have been in want of his information."

It was not merely in the acquisition of knowledge, commonly so called, that this practice contributed to prepare Farragut for his great mission as a naval commander-in-chief, but also in the discipline of character and in the development of natural capacities admirably suited for that position. It should not be overlooked that before the war, and now again in our own day, the idea of professional improvement in the United States Navy has fastened for its fitting subject upon the development of the material of war, to the comparative exclusion of the study of naval warfare. This naturally results from the national policy, which does not propose to put afloat a fleet in the proper sense of the word; and whose ideal is a number, more or less small, of cruisers neither fitted nor intended for combined action. Under these circumstances, the details of the internal economy of the single ship usurp in the professional mind an undue proportion of the attention which, in a rightly constituted navy, might far better be applied to the study of naval tactics, in the higher sense of that word, and of naval campaigns. Farragut could not but feel the influence of this tendency, so strongly marked in the service to which he belonged; the more so, as it is a thoroughly good

---

* Captain Jervis and Earl St. Vincent were the same officer under different appellations.

tendency when not pushed to an exclusive extent. But here the habit of study, and stretching in every direction his interest in matters professional, stood him in good stead, and prepared him unconsciously for destinies that could not have been foreseen. The custom of reading had made him familiar with the biography and history of his profession, the school to which the great Napoleon recommended all who would fit themselves for high military command; and of which a recent distinguished authority has said that it may be questioned whether a formulated art of war can be said to exist, except as the embodiment of the practice of great captains illustrated in their campaigns.

From these, with his great natural aptitudes for war, Farragut quickly assimilated its leading principles, which he afterward so signally illustrated in act and embodied in maxims of his own that have already been quoted. He did not employ the terminology of the art, which, though possibly pedantic in sound, is invaluable for purposes of discussion; but he expressed its leading principles in pithy, homely phrases of his own, which showed how accurate his grasp of it was. "If once you get in a soldier's rear, he is gone," was probably in part a bit of good-natured chaff at the sister profession; but it sums up in a few words the significance and strategic importance of his course in passing the batteries of the river forts, of Port Hudson and of Mobile, and brings those brilliant actions into strict conformity with the soundest principles of war. The phrases, whose frequent repetition shows how deep a hold they had taken upon him—"The more you hurt the enemy the less he will hurt you"—"The *best* protection against the enemy's fire is a well-directed fire from our own guns"—sum up one of the profoundest of all military truths, easily confessed but with difficulty lived up to, and which in these days of armor protection needs to be diligently recalled as a qualifying consideration. It is, in fact, a restatement of the oft-admitted, readily-forgotten maxim that offense is the best defense. "I believe in celerity," said he, when announcing his determination soon to pass the Mississippi forts; and good reason had he to congratulate himself that this faith showed itself in his works below New Orleans, and to lament before Mobile the failure of his Government to observe the maxim which all acknowledge. "Five minutes," said Nelson, "may make the difference between victory and defeat." "False (circuitous) routes and lost moments," wrote Napoleon, "are the determining elements of naval campaigns." All admit the value of time; but with what apathetic deliberation is often watched the flight of hours which are measuring the race between two enemies!

The personal character of Admiral Farragut afforded the firm natural foundation upon which alone a great military character can be built; for while no toleration should be shown to the absurd belief that military eminence leaps fully grown into the arena, like Minerva from the head of Jupiter—that, unlike every other kind of perfection, it grows wild and owes nothing to care, to arduous study, to constant preparation—it is still true that it can be developed only upon great natural aptitudes. The distinction conveyed by a phrase of Jomini, applied to Carnot, the great war minister of the French Revolution, is one that it is well for military and naval officers to bear constantly in mind. "Carnot," he says, although a soldier by profession, "was rather a man with a natural genius for war than an acccomplished (*instruit*) officer;" and to the lack of that studious preparation which marked Napoleon he attributes the mistakes which characterized some of Carnot's projects, although as a whole his career showed profound intuitions in the conduct of war. It is open to many able men to be accomplished and valuable officers; a few only—how few, the annals of the past show—receive the rare natural gifts which in their perfect combination make the great captain the highest manifestation of power attainable by human faculties.

The acquirements of the accomplished officer may enable him to see the right thing to be done under given conditions, and yet fail to lift him to the height of due performance. It is in the strength of purpose, in the power of rapid decision, of instant action, and, if need be, of strenuous endurance through a period of danger or of responsibility, when the terrifying alternatives of war are vibrating in the balance, that the power of a great captain mainly lies. It is in the courage to apply knowledge under conditions of exceptional danger; not merely to see the true direction for effort to take, but to dare to follow it, accepting all the risks and all the chances inseparable from war, facing all that defeat means in order thereby to secure victory if it may be had. It was upon these inborn moral qualities that reposed the conduct which led Farragut to fame. He had a clear eye for the true key of a military situation, a quick and accurate perception of the right thing to do at a critical moment, a firm grip upon the leading principles of war; but he might have had all these and yet miserably failed. He was a man of most determined will and character, ready to tread down or fight through any obstacles which stood in the path he saw fit to follow. Of this a conspicuous instance was given in the firmness with which he withstood the secession clamor of Norfolk, his outspoken defense of the unpopular Government measures, and the promptitude with which he left the place, sundering so many associations at the call of duty; and to this exhibition of strength of purpose, through the impression made upon Mr. Fox, was largely due his selection for command in the Gulf.

One of the greatest of naval commanders, whose experience of men extended through an unusually long and varied career—Earl St. Vincent—has declared that the true test of a man's courage is his power to bear responsibility; and Farragut's fearlessness of responsibility in order to accomplish necessary ends, while yet captain of a single ship, was the subject of admiring comment among his subordinates, who are not usually prone to recognize that quality in their commanders. "I have as much pleasure in running into port in a gale of wind," he wrote, "as ever a boy did in a feat of skill." The same characteristic was markedly shown under the weight of far greater issues in his determination to pass the river forts, in spite of remonstrances from his most able lieutenant, of cautious suggestions from other commanding officers, and with only the ambiguous instructions of the Navy Department to justify his action. It was not that the objections raised were trivial. They were of the most weighty and valid character, and in disregarding them Farragut showed not only the admirable insight which fastened upon the true military solution, but also the courage which dared to accept on his sole responsibility the immense risks of disaster which had to be taken.

The same moral force showed itself again, in combination with the most rapid decision and strength of purpose, when his ship was nearly thrown on shore under the batteries of Port Hudson; and yet more in the highest degree at that supreme moment of his life when, headed off from the path he had himself laid down, he led his fleet across the torpedo line in Mobile Bay. To the same quality must also be attributed the resolution to take his ships above Port Hudson, without orders, at the critical period of the campaign of 1863; and it is to be regretted in the interest of his renown that the merit of that fine decision, both in its military correctness and in the responsibility assumed, has not been more adequately appreciated. For the power to take these momentous decisions, Farragut was indebted to nature. He indeed justified them and his general course of action by good and sufficient reasons, but the reasons carried instant conviction to him because they struck a kindred chord in his breast. Speaking on one occasion of his gallant and accomplished fleet captain, Percival Drayton, he said: "Drayton does not know fear, and would fight the devil himself, but he believes in acting as if the enemy can never be caught unprepared; whereas I believe in judging him by ourselves, and my motto in action," he continued, quoting the celebrated words of Danton, "is, 'L'audace, et encore de l'audace, et toujours de l'audace.'"

With all his fearlessness and determination, severity was not one of Admiral Farragut's characteristics.

He was easily approachable, entering readily into conversation with all; and added much to the labors of his position as commanding officer by his great patience in listening to matters to which a subordinate might have attended. "His kindness was what most impressed me," says one officer who was a very young man when first reporting to him for duty. Another, who as a midshipman saw much of him, writes: "He had a winning smile and a most charming manner, and was jovial and talkative. If any officer or man had not spontaneous enthusiasm, he certainly infused it into him." Captain Drayton, who had many opportunities of observing, once said of him: "I did not believe any man could be great if he did not know how to say 'No,' but I see he can; for certainly here is a great man, and he is too kind-hearted to say 'No' in some cases where it should be said."

In person, Admiral Farragut was not above the medium size—about five feet six and a half inches high, upright in carriage, well-proportioned, alert and graceful in his movements. In early and middle life he was rather slight than heavy in frame; and it was not until the war, with the prolonged physical inactivity entailed by the river and blockade service, that he took on flesh. Up to that time his weight was not over one hundred and fifty pounds. He was very expert in all physical exercises, and retained his activity to the verge of old age. Even after his fiftieth year it was no unusual thing for him to call up some of the crew of the ship under his command and have a bout with the single-sticks. He felt great confidence in his mastery of his sword, which he invariably wore ashore; and when returning to the wharves at night, through low parts of a town where there was danger of molestation, he relied upon it to defend himself. "Any one wearing a sword," he used to say, "ought to be ashamed not to be proficient in its use."

For many years it was his habit on his birthday to go through certain physical exercises, or, as he worded it to a young officer of the fleet shortly before passing the river forts, to take a handspring; until he failed in doing this he should not, he said, feel that he was growing old. This practice he did not discontinue till after he was sixty. A junior officer of the Hartford writes: "When some of us youngsters were going through some gymnastic exercises (which he encouraged), he smilingly took hold of his left foot, by the toe of the shoe, with his right hand, and hopped his right foot through the bight without letting go." The lightness with which he clambered up the rigging of the flag-ship when entering Mobile Bay, and again over the side to see the extent of injury inflicted by the collision with the Lackawanna, sufficiently prove that up to the age of sixty-three he was capable of showing upon occasion the agility of a young man. This bodily vigor powerfully supported the energy of his mind, and carried him from daylight to dark, and from vessel to vessel of his fleet, in seasons of emergency, to see for himself that necessary work was being done without slackness; illustrating the saying attributed to Wellington, that a general was not too old when he could visit the outposts in person and on horseback.

The features of the admiral can best be realized from the admirable frontispiece. As a young man he had the sallow, swarthy complexion usually associated with his Spanish blood. His hair at the same period was dark brown, becoming in middle life almost black. In his later years he was partially bald—a misfortune attributed by him to the sunstroke from which he suffered in Tunis, and which he to some extent concealed by the arrangement of the hair. The contour of the face was oval, the cheek-bones rather prominent, until the cheeks filled out as he became fleshier during the war; the eyes hazel, nose aquiline, lips small and compressed. At no time could he have been called handsome; but his face always possessed the attraction given by animation of expression and by the ready sympathy which vividly reflected his emotions, easily stirred by whatever excited his amusement, anger, or sorrow. To conceal his feelings was to him always difficult, and, when deeply moved, impossible. The old quartermaster who lashed him in the rigging at Mobile Bay told afterward how the admiral came on deck again as the poor fellows who had been killed were being laid out on the port side of the quarter-deck. "It was the only time I ever saw the old gentleman cry," he said, "but the tears came in his eyes like a little child." A casual but close observer, who visited him on board the flag-ship in New Orleans, wrote thus: "His manners are mild and prepossessing, but there is nothing striking in his presence, and the most astute physiognomist would scarcely suspect the heroic qualities that lay concealed beneath so simple and unpretending an exterior; unless, indeed, one might chance to see him, as we did shortly afterward, just on receipt of the news from Galveston, or again on the eve of battle at Port Hudson. On such occasions the flashing eye and passionate energy of his manner revealed the spirit of the ancient vikings."

Throughout his life, from the time that as a lad still in his teens he showed to Mr. Folsom his eagerness to learn, Farragut was ever diligent in the work of self-improvement, both professional and general. His eyes were weak from youth, but he to some extent remedied this disability by employing readers in the different ships on board which he sailed; and to the day of his death he always had some book on hand. Having an excellent memory, he thus accumulated a great deal of information besides that gained from observation and intercourse with the world. Hobart Pasha, a British officer in the Turkish Navy and an accomplished seaman, wrote: "Admiral Farragut, with whom I had many conversations, was one of the most intelligent naval officers of my acquaintance." He loved an argument, and, though always good-tempered in it, was tenacious of his own convictions when he thought the facts bore out his way of interpreting their significance. When told by a phrenologist that he had an unusual amount of self-esteem, he replied: "It is true, I have; I have full confidence in myself and in my judgment"—a trait of supreme importance to a man called to high command. But against the defects of this quality he was guarded by the openness of mind which results from the effort to improve and to keep abreast of the times in which one lives.

Farragut was naturally conservative, as seamen generally tend to be; but while averse to sudden changes, and prone to look with some distrust upon new and untried weapons of war, he did not refuse them, nor did they find in him that prejudice which forbids a fair trial and rejects reasonable proof. Of ironclads and rifled guns, both which in his day were still in their infancy, he at times spoke disparagingly; but his objection appears to have arisen not from a doubt of their efficacy—the one for protection, the other for length of range—but from an opinion as to their effect upon the spirit of the service. In this there is an element of truth as well as of prejudice; for the natural tendency of the extreme effort for protection undoubtedly is to obscure the fundamental truth, which he constantly preached, that the best protection is to injure the enemy. Nor was his instinct more at fault in recognizing that the rage for material advance, though a good thing, carries with it the countervailing disposition to rely upon perfected material rather than upon accomplished warriors to decide the issue of battle. To express a fear such as Farragut's, that a particular development of the material of war would injure the tone of the service, sounds to some as the mere echo of Lever's commissary, who reasoned that the abolition of pig-tails would sap the military spirit of the nation—only that, and nothing more. It was, on the contrary, the accurate intuition of a born master of war, who feels, even without reasoning, that men are always prone to rely upon instruments rather than upon living agents—to think the armor greater than the man.

The self-confidence which Farragut exhibited in his military undertakings was not only a natural trait; it rested also upon a reasonable conviction of his mastery of his profession, resulting from long years of exclusive and sustained devotion. He did not carry the same feeling into other matters with which he had no familiarity; and he was jealously careful not to hazard the good name, which was the honor of his country as well as of himself, by attaching it to enterprises whose character he did not understand, or to duties for which he did not feel fitted. Accordingly, he refused a request made to him to allow his name to be used as director of a company, accompanied by an intimation that stock representing one hundred thousand dollars had been placed in his name on the books. "I have determined," he replied, "to decline entering into any business which I have neither the time nor perhaps the ability to attend to." In like manner he refused to allow his name to be proposed for nomination as a presidential candidate. "My entire life has been spent in the navy; by a steady perseverance and devotion to it I have been favored with success in my profession, and to risk that reputation by entering a new career at my advanced age, and that career one of which I have little or no knowledge, is more than any one has a right to expect of me."

Farragut was essentially and unaffectedly a religious man. The thoughtfulness and care with which he prepared for his greater undertakings, the courage and fixed determination to succeed with which he went into battle, were tempered and graced by a profound submission to the Almighty will. Though not obtruded on the public, his home letters evince how constantly the sense of this dependence was present to his thoughts; and he has left on record that, in the moment of greatest danger to his career, his spirit turned instinctively to God before gathering up its energies into that sublime impulse, whose lustre, as the years go by, will more and more outshine his other deeds as the crowning glory of them all—when the fiery admiral rallied his staggered column, and led it past the hostile guns and the lost Tecumseh into the harbor of Mobile.

INDEX.

THE END.

Made in the USA
Monee, IL
04 November 2026

67387507R00197